Advances in Experimental Medicine and Biology

Neuroscience and Respiration

Volume 866

More information about this series at http://www.springer.com/series/13457

Mieczyslaw Pokorski

Editor

Noncommunicable Diseases

 Springer

Editor
Mieczyslaw Pokorski
Public Higher Medical Professional School in Opole
Institute of Nursing
Opole, Poland

ISSN 0065-2598   ISSN 2214-8019   (electronic)
Advances in Experimental Medicine and Biology
ISBN 978-3-319-19973-3   ISBN 978-3-319-19974-0   (eBook)
DOI 10.1007/978-3-319-19974-0

Library of Congress Control Number: 2015946059

Springer Cham Heidelberg New York Dordrecht London

Printed on acid-free paper

Springer International Publishing AG Switzerland is part of Springer Science+Business Media
(www.springer.com)

Preface

The book series *Neuroscience and Respiration* presents contributions by expert researchers and clinicians in the field of pulmonary disorders. The chapters provide timely overviews of contentious issues or recent advances in the diagnosis, classification, and treatment of the entire range of pulmonary disorders, both acute and chronic. The texts are thought as a merger of basic and clinical research dealing with respiratory medicine, neural and chemical regulation of respiration, and the interactive relationship between respiration and other neurobiological systems such as cardiovascular function or the mind-to-body connection. The authors focus on the leading-edge therapeutic concepts, methodologies, and innovative treatments. Pharmacotherapy is always in the focus of respiratory research. The action and pharmacology of existing drugs and the development and evaluation of new agents are the heady area of research. Practical data-driven options to manage patients will be considered. New research is presented regarding older drugs, performed from a modern perspective or from a different pharmacotherapeutic angle. The introduction of new drugs and treatment approaches in both adults and children also is discussed.

Lung ventilation is ultimately driven by the brain. However, neuropsychological aspects of respiratory disorders are still mostly a matter of conjecture. After decades of misunderstanding and neglect, emotions have been rediscovered as a powerful modifier or even the probable cause of various somatic disorders. Today, the link between stress and respiratory health is undeniable. Scientists accept a powerful psychological connection that can directly affect our quality of life and health span. Psychological approaches, by decreasing stress, can play a major role in the development and therapy of respiratory diseases.

Neuromolecular aspects relating to gene polymorphism and epigenesis, involving both heritable changes in the nucleotide sequence and functionally relevant changes to the genome that do not involve a change in the nucleotide sequence, leading to respiratory disorders will also be tackled. Clinical advances stemming from molecular and biochemical research are but possible if the research findings are translated into diagnostic tools, therapeutic procedures, and education, effectively reaching physicians and patients. All these cannot be achieved without a multidisciplinary, collaborative, bench-to-bedside approach involving both researchers and clinicians.

The societal and economic burden of respiratory ailments has been on the rise worldwide, leading to disabilities and shortening of life span. COPD alone causes more than three million deaths globally each year. Concerted efforts are required to improve this situation, and part of those efforts are gaining insights into the underlying mechanisms of disease and staying abreast with the latest developments in diagnosis and treatment regimens. It is hoped that the books published in this series will assume a leading role in the field of respiratory medicine and research and will become a source of reference and inspiration for future research ideas.

I would like to express my deep gratitude to Mr. Martijn Roelandse and Ms. Tanja Koppejan from Springer's Life Sciences Department for their genuine interest in making this scientific endeavor come through and in the expert management of the production of this novel book series.

Opole, Poland   Mieczyslaw Pokorski

Volume 15: Noncommunicable Diseases

Diseases of the respiratory system often cause multisystem dysfunction and morbidity. Respiratory diseases not transmissible by a direct contact are rarer than those of inflammatory or infectious background. Such noncommunicable diseases, often entailing genetic and immune aspects, are areas of limited understanding; sarcoidosis being a case in point. This book tackles the issues relevant to such diseases. The research on novel cytokine markers, which may help in the diagnosis and management of sarcoidosis, is described. Modern approaches to the management of pneumothorax, a frequent accompaniment of lung diseases or chest wall trauma are dealt with as well. There are also chapters that underscore the immuno-inflammatory mechanisms of disorders seemingly unrelated to respiration, such as obesity or aplastic anemia, which may appreciably affect the control of the respiratory system and thus its vulnerability to diseases. The book will be of interest to clinicians and medical researchers.

Contents

Interleukin-33 as a New Marker of Pulmonary
Sarcoidosis.. 1
W. Naumnik, B. Naumnik, W. Niklińska, M. Ossolińska,
and E. Chyczewska

Finite Elements Modeling in Diagnostics of Small
Closed Pneumothorax................................. 7
J. Lorkowski, M. Mrzygłód, and O. Grzegorowska

Comparison of Small Bore Catheter Aspiration and Chest Tube
Drainage in the Management of Spontaneous Pneumothorax..... 15
P. Korczyński, K. Górska, J. Nasiłowski, R. Chazan,
and R. Krenke

Brown Adipose Tissue and Browning Agents: Irisin
and FGF21 in the Development of Obesity in Children
and Adolescents...................................... 25
B. Pyrżak, U. Demkow, and A.M. Kucharska

Regulatory T Cells in Obesity........................... 35
Anna M. Kucharska, Beata Pyrżak, and Urszula Demkow

Peroxynitrite in Sarcoidosis: Relation to Mycobacterium
Stationary Phase..................................... 41
A. Dubaniewicz, L. Kalinowski, M. Dudziak, A. Kalinowska,
and M. Singh

Effects on Lung Function of Small-Volume Conventional
Ventilation and High-Frequency Oscillatory Ventilation
in a Model of Meconium Aspiration Syndrome............... 51
L. Tomcikova Mikusiakova, H. Pistekova, P. Kosutova,
P. Mikolka, A. Calkovska, and D. Mokra

Expression of HIF-1A/VEGF/ING-4 Axis in Pulmonary
Sarcoidosis.. 61
W.J. Piotrowski, J. Kiszałkiewicz, D. Pastuszak-Lewandoska,
P. Górski, A. Antczak, M. Migdalska-Sęk, W. Górski,
K.H. Czarnecka, D. Domańska, E. Nawrot,
and E. Brzeziańska-Lasota

**Factors Influencing Utilization of Primary Health Care
Services in Patients with Chronic Respiratory Diseases** 71
D. Kurpas, M.M. Bujnowska-Fedak, A. Athanasiadou,
and B. Mroczek

**Influence of Iron Overload on Immunosuppressive
Therapy in Children with Severe Aplastic Anemia** 83
Katarzyna Pawelec, Małgorzata Salamonowicz, Anna Panasiuk,
Elżbieta Leszczynska, Maryna Krawczuk-Rybak, Urszula Demkow,
and Michał Matysiak

Index 91

Advs Exp. Medicine, Biology - Neuroscience and Respiration (2015) 15: 1–6
DOI 10.1007/5584_2015_142
© Springer International Publishing Switzerland 2015
Published online: 29 May 2015

Interleukin-33 as a New Marker of Pulmonary Sarcoidosis

W. Naumnik, B. Naumnik, W. Niklińska, M. Ossolińska,
and E. Chyczewska

Abstract

The mechanisms of sarcoidosis (Besniera-Boeck-Schaumann disease, BBS) remain incompletely understood, although recent observations suggested an important contribution of interleukin-33 (IL-33). So far, there are no data about bronchoalveolar lavage fluid (BALF) concentration of IL-33 in patients with BBS. In the present study we attempted to relate the concentration of IL-33 to IL-18, a well-known marker of BBS activity, in BALF of BBS patients. We examined 24 BBS patients (stage II). The age-matched control group consisted of 24 healthy subjects. The levels of IL-33 and IL-18 in BALF were higher in BBS patients than in the control group [IL-33: 4.8 (0.1–12.5) $vs.$ 3.4 (0.6–56.9) pg/ml, p = 0.024; IL-18: 33.2 (5.7–122.0) $vs.$ 10.8 (1.9–45.8) pg/ml, p = 0.002]. In the BBS group, the correlations between IL-33 and IL-18 (r = 0.606, p = 0.002), and between IL-33 and diffusion lung capacity for carbon monoxide (DLCO) (r = −0.500, p = 0.035) were found. The receiver-operating characteristic curves were applied to find the cut-off serum levels of IL-33 and IL-18 in BALF (BBS $vs.$ healthy: IL-33 2.7 pg/ml and IL-18 16.4 pg/ml). We conclude that IL-33 appears an important factor of pulmonary BBS activity.

Keywords

Bronchoalveolar lavage fluid • Interleukin-33 • Interleukin-18 • Disease marker • Sarcoidosis

W. Naumnik (✉)
Department of Lung Diseases, Medical University of Bialystok, 14 Zurawia St., PL 15-540 Bialystok, Poland

Department of Clinical Molecular Biology, Medical University of Bialystok, Bialystok, Poland
e-mail: wojciechnaumnik@gmail.com

B. Naumnik
First Department of Nephrology and Transplantation with Dialysis Unit, Medical University of Bialystok, Bialystok, Poland

W. Niklińska
Department of Histology and Embryology, Medical University of Bialystok, Bialystok, Poland

M. Ossolińska and E. Chyczewska
Department of Lung Diseases, Medical University of Bialystok, 14 Zurawia St., PL 15-540 Bialystok, Poland

1   Introduction

Sarcoidosis is a multisystem immunologic disorder of unknown etiology, characterized initially by a Th (T-helper) lymphocyte/macrophage alveolitis and later by the formation of epithelioid cell granulomas (Cui et al. 2010). The mechanisms of sarcoidosis (Besniera-Boeck-Schaumann disease, BBS), the most common interstitial lung disease (ILD), remain incompletely understood, although recent observations have suggested an important contribution of interleukin 33 (IL-33) (Luzina et al. 2013). IL-33 can be classified as an alarmin because it is released into the extracellular space following cell damage or tissue injury and acts as an endogenous danger signal by sending out warning signals to alert neighbouring cells and tissues (Liew et al. 2010). IL-33 was thought to be mainly involved in initiating and perpetuating Th2-driven responses and activating mast cells, due to the fact that these two cell types express high amounts of ST2 (the interleukin-1 receptor family member), the cell-surface receptor of IL-33 (Schmitz et al. 2005). Recently many studies have revealed that IL-33 is associated with suppressing of Th1 responses, which plays a major role in patients with sarcoidosis (Rostan et al. 2013; Smithgall et al. 2008). Moreover, IL-33 is involved in the control of cell cycle progression in endothelial cell and regulation of angiogenesis (Martin 2013). Li et al. (2014) showed that IL-33 promotes lung fibrosis in mice. Some patients with BBS can develop lung fibrosis at a later stage of disease (Cui et al. 2010). Angiogenesis may contribute to fibroproliferation, which may lead to novel therapeutic options. According to some authors there is an association between interleukin 18 (IL-18) and activity of pulmonary sarcoidosis (Liu et al. 2010). There have been no data about concentrations of IL-33 in patients with sarcoidosis so far. Therefore, in the present study we aimed to compare the concentration of IL-33 to IL-18.

2   Methods

The study was conducted in conformity with the Declaration of Helsinki for Human Experimentation of the World Medical Association. The protocol was approved by a local Ethics Committee and written informed consent was obtained from all participants.

2.1   Patients and Control Subjects

The study group included 24 patients with lung sarcoidosis (BBS). Patients were in stage 2 of BBS (bilateral hilar lymphadenopathy and pulmonary infiltrations: F/M - 5/19, mean age 40 ± 8 years) recruited at the Department of Lung Diseases, the Medical University of Białystok in Poland, in 2009–2013. We diagnosed patients according to the current clinical and pathological guidelines (Statement on sarcoidosis 1999). The control group consisted of 24 healthy volunteers (F/M - 5/19, mean age 39 ± 9 years) without any inflammatory conditions. All patients and control subjects underwent BALF and lung function tests (vital capacity, VC, and diffusing capacity for carbon monoxide, DLCO) (Standardization of Spirometry 1994 Update. American Thoracic Society 1995). We performed bronchofiberoscopy with BALF as part of a routine clinical management. We used fiberoptic bronchoscope (Pentax FB 18 V, Pentax Corporation, Tokyo, Japan) under local anesthesia with lidocaine, following premedication with intramuscular atropine and hydroxyzine as a sedative. The bronchoscope was inserted and wedged in the right middle lobe, and three 50 ml aliquots of sterile saline solution, warmed to 37 °C, were instilled and recovered from the subsegmental bronchus by suction. The recovered fluid was filtered through 2 layers of sterile gauze and subsequently centrifuged at 800 rpm for 10 min at 4 °C. Supernatant was stored at -70 °C until use. BALF samples were analyzed for total and differential cell counts, flow cytometry to measure CD4+, and CD8+ lymphocyte counts, and for IL-33 and IL-18 detected by Elisa. Cell differentials were made on smears stained by Grünwald-Giemza by counting at least 400 cells under a light microscope (magnification $\times$ 1000). Another part of the BALF (cell suspension) was incubated with phycoerythrin-labeled anti-CD4 antibody and fluorescein isothiocyanate-labeled anti-CD8 antibody (Becton Dickinson, Mountain View, CA) for 20 min, washed twice. Flow cytometry was performed using Becton Dickinson flow cytometer that detects lymphocytes by fluorescence. The percentages of positively stained cells were scored to determine the number of CD4 and CD8 cells.

2.2 Concentrations of IL-33 and IL-18 in BALF

Concentrations of IL-33 were measured by commercially available enzyme-linked immunosorbent assays (ELISA) (eBioscience, San Diego, CA) and (MBL International, Woburn, MA, respectively). The assays were performed according to manufacturers' recommendations. The minimum detectable levels of IL-33 and IL-18 were 0.2 pg/mL and 1.5 pg/ml, respectively.

2.3 Statistical Analysis

The Shapiro-Wilk test was used for data distribution analysis. We used a t-test to calculate the parametrical data. The Mann-Whitney U and Wilcoxon tests were used for the features inconsistent with normal data distribution. Spearman's rank test was used to calculate correlations between the parameters and receiver-operating characteristics (ROC) curves to find the cut-off levels of IL-33 and IL-18. A value of $p < 0.05$ was considered to indicate statistical significance. We used Statistica 10.0 software (StatSoft Inc., Tulsa, USA) for all analyses.

3 Results

There were no appreciable differences in age or gender between the patient and control groups. Concerning pulmonary function, BBS patients had lower %VC and %DLCO than those in healthy persons (%VC: 82.1 ± 21.2 vs.

96.4 ± 6.0, p = 0.02; %DLCO: 80.1 ± 25.0 vs. 95.3 ± 9.0, p = 0.01).

The levels of IL-33 and IL-18 in BALF were higher in the BBS patients than those in the control group [IL-33: 4.8 (0.1–12.5) pg/ml vs. 3.4 (0.6–56.9) pg/ml, p = 0.024; IL-18: 33.2 (5.7–122.0) pg/ml vs. 10.8 (1.9–45.8) pg/ml, p = 0.004] (Fig. 1a, b). There was a positive correlation between BALF levels of IL-33 and IL-18 in the BBS group (r = 0.606, p = 0.002) (Fig. 2).

The ROC curves show that specificity and sensitivity of BALF IL-33 in the BBS patients in relation to healthy people were 59 % and 95 %, respectively, at a cut-off value of 2.699 pg/ml. Specificity and sensitivity of BALF IL-18 in the BBS patients in relation to healthy people were 28 % and 80 %, respectively, at a cut-off value of 16.436 pg/ml. The areas under the curve for IL-33 and IL-18 in BALF were 0.683 and 0.781, respectively (Fig. 3).

In the BBS group, DLCO% correlated negatively with IL-33 (r = −0.512, p = 0.035) (Fig. 4a), and IL-18 (r = −0.596, p = 0.009; not shown) in BALF. Moreover, a positive correlation was found between IL-33 and the percentage of lymphocytes (r = 0.45, p = 0.032) in BALF (Fig. 4b).

The percentage of lymphocytes was higher, and that of macrophages was lower in BALF of BBS patients compared with those in the healthy group (%lymphocytes: 46.7 ± 28.0 vs. 16.1 ± 7.0, p = 0.002; %macrophages: 59.4 ± 23.1 vs. 82.1 ± 15.0, p = 0.003). The BBS patients had a higher percentage of CD4+ in BALF than healthy subjects (%CD4+: 48.4 ± 13.2 vs. 8.1 ± 0.3, p = 0.003). There were no significant differences between the percentage of CD8+ in BALF of BBS and control groups (CD8+: 15.3 ± 3.1 vs. 16.8 ± 4.2, p = 0.422).

4 Discussion

Mortality in patients with sarcoidosis is higher than that in the general population, due mainly to pulmonary fibrosis (Valeyre et al. 2014). The

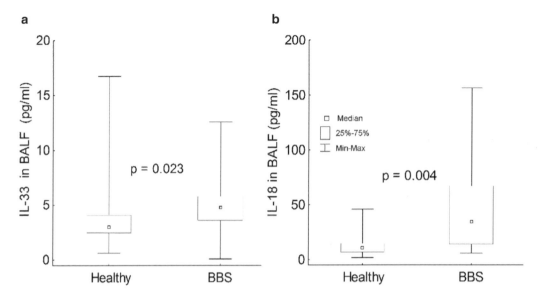

Fig. 1 IL-33 (a) and IL-18 (b) concentrations in BALF from sarcoidosis (BBS) patients and control subjects

Fig. 2 **Correlation**
between IL-33 and IL-18
concentrations in BALF
from sarcoidosis (BBS)
patients

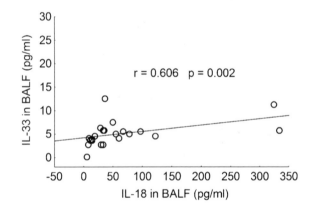

finding of the underlying mechanism of sarcoidosis and the elucidation of relevant biomarkers is key for future therapeutic advances. In the resent study, BBS patients had higher levels of IL-33 than those in healthy subjects. This observation is consistent with the study of Yang et al. (2011). Those authors described that IL-33 can contribute to the development of Th1-type of immune response as well as enhanced IL-18 secretion. There are many publications about Th1 immune responses in BBS, substantially less is known about cellular sources of IL-33. We found a correlation between the level of IL-33 and the percentage of lymphocytes in BALF from BBS patients. That may suggest that IL-33 is released from lymphocytes in these patients. The result is in line with a study of Zissel et al. (2010) who showed that an exaggerated immune response stemming from the close interaction between macrophages and T cells causes the formation of sarcoid lesions. Our observations are not in accord with a study of Baekkevold et al. (2003) who described that IL-33 is released by endothelial and epithelial cells, but are consistent with a study of Kempf et al. (2014) who demonstrated that granulomas are a source of interleukin-33 expression in pulmonary and extrapulmonary sarcoidosis. Summarising, it is possible that IL-33 is released from both lymphocytes and epithelial cells. Epithelioid-cell-rich granulomas and stimulation of T-cells plays a key role in the pathogenesis of sarcoidosis (Chen et al. 2011). IL-33 is released from stressed or damaged cells to the tissue microenvironment. This cytokine stimulates the proinflammatory interferon gamma and profibrotic factors such as transforming growth factor, insulin-like growth factor, and matrix metalloproteinases. All these factors are present at sites of inflammation in individuals with BBS (Kempf et al. 2014; Kakkar et al. 2012).

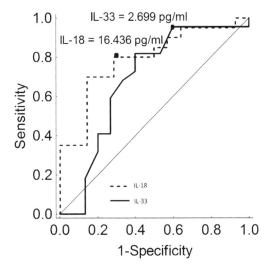

Fig. 3 Receiver operating characteristic (ROC) curve for IL-33 and IL-18 in BALF differentiating sarcoidosis (BBS) and healthy subjects (AUC 0.683 and 0.781, respectively)

IL-18 expression is increased in airway epithelial cells in active BBS (Kieszko et al. 2007). That finding is in accord with the present study in which BBS patients had a higher level of IL-18 in BALF than that in healthy subjects. IL-18 is defined as an inducer of Th1 response. This cytokine is produced by macrophages, dendritic cells, and airway epithelial cells (Okamura et al. 1995). IL-18 is considered as a good marker of BBS activity (Kieszko et al. 2007). In the present study, we found a strong correlation between the concentration of IL-33 and IL-18 in BALF from BBS patients. Moreover, the level of IL-33 correlated with DLCO. These results are consistent with the findings of Li et al. (2014) who showed that IL-33 could promote, enhancing the production of profibrotic factors, the process of chronic inflammation and fibroproliferation, contributing to pulmonary fibrosis and lung dysfunction. In order to fully verify this issue, sarcoidosis patients should be investigated at sequential stages of the disease. Our present patients were at stage II of BBS, without lung fibrosis. Liu et al. (2011) postulate that the measurement of circulating IL-18 might be of potential clinical utility in the differential diagnosis of BBS *versus* idiopatic pulmonary fibrosis.

We surmise that IL-33 may be a potent marker of BBS activity that can be used in clinical practice. A recent study of Kempf et al. (2014)

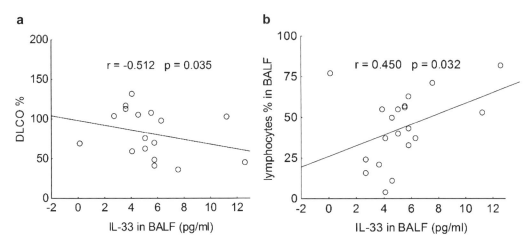

Fig. 4 Correlations between IL-33 and (a) DLCO and (b) %lymphocytes in BALF from sarcoidosis (BBS) patients

reported that IL-33 plays a critical role in the pathogenesis and progression of sarcoidosis. IL-33 expression in sarcoidosis seems to depend on the specific tissue microenvironment of sarcoid granulomas and represents a novel biomarker for systemic involvement. In summary, our present findings point to the possibility of practical use of IL-33 measurements. That could contribute to a more individualized treatment of patients.

Conflicts of Interest The authors had no conflicts of interest to declare in relation to this article.

References

American Thoracic Society. Single-breath carbon monoxide diffusing capacity (transfer factor). Recommendations for a standard technique–1995 update (1995) Am J Respir Crit Care Med 152:2185–2198

Baekkevold ES, Roussigné M, Yamanaka T, Johansen FE, Jahnsen FL, Amalric F, Brandtzaeg P, Erard M, Haraldsen G, Girard JP (2003) Molecular characterization of NF-HEV, a nuclear factor preferentially expressed in human high endothelial venules. Am J Pathol 163:69–79

Chen ES, White ES (2011) Innate pathways shape sarcoidosis signaling: from bugs to drugs. Am J Respir Crit Care Med 183:425–427

Cui A, Anhenn O, Theegarten D, Ohshimo S, Bonella F, Sixt SU, Peters J, Sarria R, Guzman J, Costabel U (2010) Angiogenic and angiostatic chemokines in idiopathic pulmonary fibrosis and granulomatous lung disease. Respiration 80:372–378

Kakkar R, Hei H, Dobner S, Lee RT (2012) Interleukin 33 as a mechanically responsive cytokine secreted by living cells. J Biol Chem 287:6941–6948

Kempf W, Zollinger T, Sachs M, Ullmer E, Cathomas G, Dirnhofer S, Mertz KD (2014) Granulomas are a source of interleukin-33 expression in pulmonary and extrapulmonary sarcoidosis. Hum Pathol 45:2202–2210

Kieszko R, Krawczyk P, Jankowska O, Chocholska S, Król A, Milanowski J (2007) The clinical significance of interleukin 18 assessment in sarcoidosis patients. Respir Med 101:722–728

Li D, Guabiraba R, Besnard AG, Komai-Koma M, Jabir MS, Zhang L, Graham GJ, Kurowska-Stolarska M, Liew FY, McSharry C, Xu D (2014) IL-33 promotes ST2-dependent lung fibrosis by the induction of alternatively activated macrophages and innate lymphoid cells in mice. J Allergy Clin Immunol 134:1422–1432

Liew FY, Pitman NI, McInnes IB (2010) Disease-associated functions of IL-33: the new kid in the IL-1 family. Nat Rev Immunol 10:103–110

Liu DH, Yao YT, Cui W, Chen K (2010) The association between interleukin-18 and pulmonary sarcoidosis: a meta-analysis. Scand J Clin Lab Invest 70:428–432

Liu DH, Cui W, Chen Q, Huang CM (2011) Can circulating interleukin-18 differentiate between sarcoidosis and idiopathic pulmonary fibrosis? Scand J Clin Lab Invest 71:593–597

Luzina IG, Kopach P, Lockatell V, Kang PH, Nagarsekar A, Burke AP, Hasday JD, Todd NW, Atamas SP (2013) Interleukin-33 potentiates bleomycin-induced lung injury. Am J Respir Cell Mol Biol 49:999–1008

Martin MU (2013) Special aspects of interleukin-33 and the IL-33 receptor complex. Semin Immunol 25:449–457

Okamura H, Tsutsi H, Komatsu T, Yutsudo M, Hakura A, Tanimoto T, Torigoe K, Okura T, Nukada Y, Hattori K (1995) Cloning of a new cytokine that induces IFN-gamma production by T cells. Nature 378:88–91

Rostan O, Gangneux JP, Piquet-Pellorce C, Manuel C, McKenzie AN, Guiguen C, Samson M, Robert-Gangneux F (2013) The IL-33/ST2 axis is associated with human visceral leishmaniasis and suppresses Th1 responses in the livers of BALB/c mice infected with Leishmania donovani. MBio 17:e00383–13

Schmitz J, Owyang A, Oldham E, Song Y, Murphy E, McClanahan TK, Zurawski G, Moshrefi M, Qin J, Li X, Gorman DM, Bazan JF, Kastelein RA (2005) IL-33, an interleukin-1-like cytokine that signals via the IL-1 receptor-related protein ST2 and induces T helper type 2-associated cytokines. Immunity 23:479–490

Smithgall MD, Comeau MR, Yoon BR, Kaufman D, Armitage R, Smith DE (2008) IL-33 amplifies both Th1- and Th2-type responses through its activity on human basophils, allergen-reactive Th2 cells, iNKT and NK cells. Int Immunol 20:1019–1030

Standardization of Spirometry, 1994 Update. American Thoracic Society (1995) Am J Respir Crit Care Med 152:1107–1136

Statement on sarcoidosis. Joint Statement of the American Thoracic Society (ATS), the European Respiratory Society (ERS) and the World Association of Sarcoidosis and Other Granulomatous Disorders (WASOG) adopted by the ATS Board of Directors and by the ERS Executive Committee (1999) Am J Respir Crit Care Med 160:736–755

Valeyre D, Prasse A, Nunes H, Uzunhan Y, Brillet PY, Müller-Quernheim J (2014) Sarcoidosis. Lancet 383:1155–1167

Yang Q, Li G, Zhu Y, Liu L, Chen E, Turnquist H, Zhang X, Finn OJ, Chen X, Lu B (2011) IL-33 synergizes with TCR and IL-12 signaling to promote the effector function of CD8+ T cells. Eur J Immunol 11:3351–3360

Zissel G, Prasse A, Müller-Quernheim J (2010) Immunologic response of sarcoidosis. Semin Respir Crit Care Med 31:390–403

Advs Exp. Medicine, Biology - Neuroscience and Respiration (2015) 15: 7–13
DOI 10.1007/5584_2015_150
© Springer International Publishing Switzerland 2015
Published online: 29 May 2015

Finite Elements Modeling in Diagnostics of Small Closed Pneumothorax

J. Lorkowski, M. Mrzygłód, and O. Grzegorowska

Abstract

Posttraumatic pneumothorax still remains to be a serious clinical problem and requires a comprehensive diagnostic and monitoring during treatment. The aim of this paper is to present a computer method of modeling of small closed pneumothorax. Radiological images of 34 patients of both sexes with small closed pneumothorax were taken into consideration. The control group consisted of X-rays of 22 patients treated because of tension pneumothorax. In every single case the model was correlated with the clinical manifestations. The procedure of computational rapid analysis (CRA) for *in silico* analysis of surgical intervention was introduced. It included implementation of computerize tomography images and their automatic conversion into 3D finite elements model (FEM). In order to segmentize the 3D model, an intelligent procedure of domain recognition was used. In the final step, a computer simulation project of fluid-structure interaction was built, using the ANSYS\Workbench environment of multi-physics analysis. The FEM model and computer simulation project were employed in the analysis in order to optimize surgical intervention. The model worked out well and was compatible with the clinical manifestations of pneumothorax. We conclude that the created FEM model is a promising tool for facilitation of diagnostic procedures and prognosis of treatment in the case of small closed pneumothorax.

Keywords

Artificial intelligence • Computer modeling • Diagnostic • Finite elements method • In silico analysis • Pneumothorax

J. Lorkowski (✉)
Department of Orthopedics and Traumatology, Central Clinical Hospital of Ministry of Interior, 137 Wołoska St., 02-507 Warsaw, Poland

Rehabilitation Center 'Health', Cracow, Poland
e-mail: jacek.lorkowski@gmail.com

M. Mrzygłód
Institute of Rail Vehicles, Faculty of Mechanical Engineering, Cracow University of Technology, Cracow, Poland

O. Grzegorowska
Rehabilitation Center 'Health', Cracow, Poland

1   Introduction

Pneumothorax means the presence of the air between parietal and visceral pleura, which entails lung collapse. As a consequence, there are a mediastinal shift to the opposite side, hemodynamic imbalance due to disordered perfusion/ventilation ratio, and hypoxia. Pneumothorax symptoms can be divided into early and late ones. The former are anxiety, tachypnea, respiratory distress, increased breathing effort, shock (hypotension and tachycardia), hyper-resonance and decreased breath sounds on the ipsilateral side, emphysema, and engorged neck veins if there is no hypovolemia. The latter appear as cyanosis and a shift of trachea to the opposite side. Cardiorespiratory failure and sudden cardiac arrest also can be observed, especially in case of tension pneumothorax. Tension pneumothorax is a serious clinical problem, which requires accurate and, quite often, rapid diagnostics and monitoring. One of the causes of tension pneumothorax is trauma. We distinguish closed, open, small closed, and tension type of pneumothorax, which can occur immediately or with some delayed after trauma (Plourde et al. 2014). It is known that mild general trauma involves the chest in about 12 %, while major trauma is accompanied by chest trauma in 47 % (Lorkowski et al. 2014a, b). Tension pneumothorax appears in about 5 % of major trauma victims in the pre-hospital setting and in 1–3 % of patients in intensive care units (Roberts et al. 2014). The small closed pneumothorax caries the lowest risk for patients (Lorkowski et al. 2013; Ball et al. 2005).

A problematic issue concerning pneumothorax is the choice of an optimal method for its diagnostic and treatment. In case of small pneumothorax, up to 15 % of lung volume or size less than 1.5 cm, conservative treatment and monitoring are advocated. Our earlier study shows that conservative treatment may possibly be extended to pneumothorax the size of up 2 cm (Lorkowski et al. 2013). A multicenter study performed in 569 blunt trauma patients shows that occult pneumothorax can be treated conservatively

when linked with monitoring. An increase in the pneumothorax size or ongoing signs of respiratory failure necessitate the undertaking of invasive approach, such as decompression and drainage (Moore et al. 2011). Concerning the diagnostic methods, the choice is the following: auscultation and percussion, ultrasonography, radiography, and computer tomography. The ultrasound effectiveness depends much on the operator's skills and experience, but it shows higher sensitivity compared with radiography (Alrajab et al. 2013; Alrajhi et al. 2012; Ding et al. 2011). The value of the ultrasound examination has been appreciated in emergency in case of major trauma patients (Ianniello et al. 2014) and in positive-pressure ventilated patients (Oveland et al. 2013). Nevertheless, ultrasonography has limited diagnostic power in case of subcutaneous emphysema, obesity, adhesive pleura diseases, or emphysema (Kreuter and Mathis 2014). Among new methods developed for the diagnosis of pneumothorax, the measurements or computer analysis of pulmonary acoustic transmission, the latter connected with visualization are of note, but do not settle all the diagnostic difficulties involved (Hayashi 2011; Mansy et al. 2002a, b). Therefore, the aim of the present paper was to present a new method of modeling of small closed pneumothorax based on *in silico* mechanical-fluid analysis using fast 3D finite elements modeling (FEM).

2   Methods

The study was accepted by an institutional Review Board for Human Research and was performed in accord with the Declaration of Helsinki for Medical Research Involving Human Subjects. Radiological images of 34 patients, F/M – 11/23, mean age 48.5 years (range 20–84 years), with small closed pneumothorax were taken into consideration. The pneumothorax size was assessed on the basis of widely accepted criteria as a distance between visceral and parietal pleura, which in the anterior-posterior radiographic projection ranged from 0.5 to 2.0 cm (Henry et al. 2003). The

Table 1 Sequential steps of computational rapid analysis (CRA) using 3D finite elements modeling (FEM)

1	Computed tomography (CT) scanning
2	Conversion of CT images into 3D FEM model
3	Assignment of material properties to greyscale shades of bitmaps
4	Use of artificial intelligence analysis to the 3D model segmentation
5	Assignment of equivalent material properties to designated subdivisions, using a multiscale approach (for each elementary volume one finite element of averaged mechanical properties was assigned).
6	FEM simulation: static structural, computational fluid dynamics (CFD), fluid-structure interaction (FSI), multi-body simulation (MBS), explicit dynamics analysis, remodeling/healing simulation.

patients were treated conservatively with no pleural drainage. The results of treatment were satisfactory, with decreasing size of pneumothorax in consecutive radiograms, and its complete resolution in each case after a month.

The control group consisted of 22 patients, F/M – 5/17, mean 48.8 years (range 21–85 years). Those were trauma patients with tension pneumothorax at the time of admission. Pneumothorax was treated with pleural drainage in this group and the course of treatment was monitored radiologically. Normalization of blood gas content and other laboratory tests was present in 20 cases in which pneumothorax was resolved and the symptoms of lung contusion disappeared after a month. Two of these patients died during treatment of polytrauma, but pneumothorax was not the cause of death in either case.

Chest radiograms of all patients were analyzed. Based on computer tomography results, a reference computer model was made. In each case, the reference model was correlated with the clinical presentation of pneumothorax. The procedural elaboration consisted of an algorithm called 'computational rapid analysis' (CRA) for *in silico* analysis and optimization of surgical intervention. The sequential procedural steps are shown in Table 1.

Computed tomography (CT) chest images of the patients were taken. Then, a set of CT images was processed into an FEM model. Considering the size of the model derived (~20 GB of database file), only a representative model part was chosen for further simulation (Fig. 1a). This fragment underwent a further processing to decrease and optimize the database size. Finally, the model was limited to its minimal volume,

covering the following subdivisions: chest cavity with the lung and the pneumothorax area, and a layer of surrounding tissues (Fig. 1b). In order to separate those divisions, an artificial intelligence (AI)-based filtration procedure was used. The procedure enabled to identify subdomains' boundaries and to assign equivalent material properties (Fig. 1c). Afterward, an attempt was made to transfer the database into a multiphysics simulation project of ANSYS/Workbench environment. Unfortunately, the model size made this procedure unworkable. The procedure turned out to be time-consuming and, with the available equipment base, could not be effectively implemented (10 % of single subdivision preprocessing required a time longer than 4 h). Therefore, an alternative model in the CATIA V5 program was prepared (Fig. 2a). Then, a simplified geometric 3D model of the investigated tissues was loaded into the computerized fluid dynamic (CFD) flow analysis of the ANSYS WB program (Fig. 2b).

3   Results

The analyzed fluid, designated as the air of temperature of 25 °C, was assumed for further analysis. After defining the boundary conditions and loads of the analysis (Fig. 3a, b), dynamic calculations of the fluid (air) flow in the chest cavity were made. This enabled to determine the distribution of pressure acting on the lung and the chest walls (Fig. 3c). Transferring the results of CFD analysis to the structural analysis environment allowed making the next methodological step, which was to create a conjugated fluid-structure analysis (FSI) (Fig. 4a). For the

Fig. 1 Preprocessing of finite elements modeling (FEM) model of small closed pneumothorax using computed tomography (CT) images: (a) isometric view of FEM model after automatic loading; (b) the model view after database optimization and removing of unnecessary finite elements; (c) the model divided into three subdivisions after applying homogenization of material properties

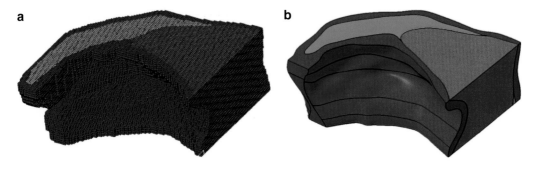

Fig. 2 A simplified geometric model of small closed pneumothorax based on computed tomography (CT) images: (a) model on the basis of CATIA V5 program; (b) model on the basis of computerized fluid dynamic (CFD) flow analysis of ANSYS WB program

structural FEM model, two kinds of homogeneous material were assumed. For the lungs, it was Young's modulus $E_l = 1$ MPa and Poisson's ratio $\nu_1 = 0.38$ (Li et al. 2013). For the tissues surrounding the chest cavity, it was $E_2 = 500$ MPa and $\nu_2 = 0.3$. On that basis, investigations of the pressure influence on the lung tissue and chest cavity deformation (Fig. 4b) (i.e., the effects of increasing the small closed pneumothorax and its conversion into tension pneumothorax) were calculated.

As a result of the dynamic FEM model created we could visualize the qualitative interrelation between pleural cavity and the lung of the patient as shown in Figs. 3 and 4. The model enabled the express observation of how the lung tissue would change depending on increased or decreased pressure in pleural cavity. In case of the patients with small closed pneumothorax and pressures between -1 and -6 mmHg for the pneumothorax size between 0.5 and 2.0 cm (an average of 1.2 cm), the model represents a static picture of non-growing pneumothorax. The mapping of deformations occurring on the lung surface in patients with small closed pneumothorax is shown in Fig. 3c, with the red color pointing to the areas of maximal pressure loads. When the negative pressure increases, the pneumothorax size decreases. The introduction of positive pressures in pleural cavity results in creating a representation of tension pneumothorax as shown in Fig. 4b. In this case, the model replicates the shape of collapsed lung due to pressure tension exerted on lung surface; the picture that complied with the X-rays taken in the patients from the control group having tension pneumothorax.

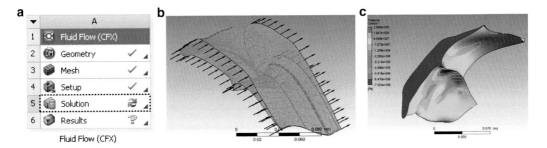

Fig. 3 Fluid-structure interaction (FSI) analysis – Part I: (**a**) computerized fluid dynamic (CFD) flow – simulation analysis; (**b**) finite elements model (FEM) with loads; (**c**) analysis result – distribution of pressure acting on the chest walls; red areas show the sites of maximal pressure loads

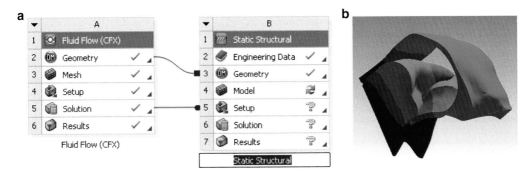

Fig. 4 Fluid-structure interaction (FSI) analysis – Part II: (**a**) computer simulation process; (**b**) analysis result – lung deformation under loads from computerized fluid dynamic (CFD) flow analysis

4 Discussion

The chest trauma is one of the most notable components of polytrauma. It is also estimated that the incidence of isolated chest trauma runs in the third place behind the head and leg trauma (Lorkowski et al. 2014b). Chest trauma most frequently involves the development of pneumothorax. When it occurs, there is a communication between the intrapulmonary air space and pleural cavity. This mechanism leads to decreasing pressure differences between pleural cavity and the lung until there is no longer pressure difference, or until the communication between these two structures is sealed (Roberts et al. 2014). The main methods that are used for the pneumothorax diagnostics are: anamnesis and physical and radiological examinations. A specific pneumothorax symptom is pain, reported in anamnesis, found on the affected side during physical examination, and strengthening during respiratory movements of the chest wall. Pneumothorax may be an underlying reason of respiratory failure that develops particularly in case of major trauma and also in case of tension pneumothorax and that calls for immediate lifesaving cardiorespiratory interventions. In fact, in the present investigation we observed respiratory failure in all but one patient with tension pneumothorax in the control group. In the group of patients with the small closed pneumothorax we did not observe the symptoms of respiratory failure, but each patient reported pain in the region of fractured ribs or damaged chest wall. Occasionally, there was breathing asymmetry. Some patients suffered from subcutaneous emphysema, which is a frequent accompaniment of closed pneumothorax (Hwang et al. 1996; Myers et al. 2002).

The FEM model presented in this article differs from other standard descriptive methods of pneumothorax assessment. With X-ray images it is impossible to foresee which lung areas will become subject to increased pressures acting on tissue. These areas are where the possible secondary, local lung contusion occurs. The distribution of forces acting on lung tissue changes in case of pneumothorax, and a lung contusion can be expected. The *in silico* model developed enables the identification of the contusion area, which makes it obviously an advantageous and more sensitive method compared with other imaging techniques. The model, jointly with the FSI analysis, apart from its already tested clinical applicability, can be used for future studies on the influence of negative pressure on the degree of lung expansion, the assessment of lung deformation, the description of pressure distribution in thorax cavity, or the prediction of pneumothorax development under the action of pressure. After substituting an acceptable level of pressure, it is feasible to set the tension limit for lung collapse, which may help predict the appearance of tension pneumothorax. The insertion of geometric features, e.g., an aperture, would enable to choose the optimal point for the chest wall puncture.

The *in silico* methodology presented above is currently used in orthopedics and traumatology (Lorkowski et al. 2012, 2013, 2014a), in radiotherapy planning in patients with liver cancer or metastases (Velec et al. 2011), in lung and esophagus cancer (Li et al. 2013), or for the purpose of injury simulations (Digges et al. 2014; Valero et al. 2014). In case of neoplasm and radiotherapy planning, the method makes the treatment more effective and helps protect surrounding tissues from excessive radiation dose. In orthopedics and traumatology, the method helps to choose the express anatomical structures for treatment simulation and optimizes the planned treatment paradigm. The assessment of dynamic fluid-structure interaction also enables to set the fluid pressure values which would have a therapeutic effect on the patient.

In conclusion, the created FEM *in silico* model gives a dynamically-oriented insight into the formation of small closed pneumothorax and may help prevent its transition into a life threatening tension pneumothorax. The model seems also of high diagnostic and treatment potential in

pulmonary disorders, particularly those which are involved with chest and lung deformations, be it diseased or traumatic in origin. The medical usefulness of FEM *in silico* analysis, which uses mathematical models of biological systems, warrants further investigative explorations. The investigations go in harmony with the ever increasing computational power, which allows for previously unattainable modeling of complex systems.

Conflicts of Interest The authors declare no conflicts of interest in relation to this article.

References

Alrajab S, Youssef AM, Akkus NI, Caldito G (2013) Pleural ultrasonography versus chest radiography for the diagnosis of pneumothorax: review of the literature and meta-analysis. Crit Care 17(5):R208

Alrajhi K, Woo MY, Vaillancourt C (2012) Test characteristics of ultrasonography for the detection of pneumothorax: a systematic review and meta-analysis. Chest 141(3):703–708

Ball CG, Kirkpatrick AW, Laupland KB, Fox DI, Nicolaou S, Anderson IB, Hameed SM, Kortbeek JB, Mulloy RR, Litvinchuk S, Boulanger BR (2005) Incidence, risk factors, and outcomes for occult pneumothoraces in victims of major trauma. J Trauma 59(4):917–925

Digges K, Eigen A, Tahan F, Grzebieta R (2014) Factors that influence chest injuries in rollovers. Traffic Inj Prev 15(Suppl 1):42–48

Ding W, Shen Y, Yang J, He X, Zhang M (2011) Diagnosis of pneumothorax by radiography and ultrasonography: a meta-analysis. Chest 140(4):859–866

Hayashi N (2011) Detection of pneumothorax visualized by computer analysis of bilateral respiratory sounds. Yonago Acta Med 54(4):75–82

Henry M, Arnold T, Harvey J (2003) BTS guidelines for the management of spontaneous pneumothorax. Thorax 58(Suppl 2):ii39–ii52

Hwang JCF, Hanowell LH, Grande CM (1996) Perioperative concerns in thoracic trauma. Baillières Clin Anaesthesiol 10(1):123–153

Ianniello S, Di Giacomo V, Sessa B, Miele V (2014) First-line sonographic diagnosis of pneumothorax in major trauma: accuracy of e-FAST and comparison with multidetector computed tomography. Radiol Med 119(9):674–680

Kreuter M, Mathis G (2014) Emergency ultrasound of the chest. Respiration 87(2):89–97

Li M, Castillo E, Zheng XL, Luo HY, Castillo R, Wu Y, Guerrero T (2013) Modeling lung deformation: a combined deformable image registration method with spatially varying Young's modulus estimates. Med Phys 40(8):081902. doi:10.1118/1.4812419

Lorkowski J, Mrzygłód M, Hładki W (2012) Phenomenon of remodeling and adjustment the topology of the calcaneus with a solitary cyst- case report. Przegl Lek 69:201–204

Lorkowski J, Teul I, Hładki W, Kotela I (2013) The evaluation of the treatment results in patients with small closed pneumothorax. Ann Acad Med Stetin 59:43–47

Lorkowski J, Mrzygłód M, Kotela A, Kotela I (2014a) Application of rapid computer modeling in the analysis of the stabilization method in intraoperative femoral bone shaft fracture during revision hip arthroplasty – a case report. Pol Orthop Traumatol 79:138–144

Lorkowski J, Teul I, Hładki W, Kotela I (2014b) The evaluation of procedure and treatment outcome in patients with tension pneumothorax. Ann Acad Med Stetin 60(1):18–23

Mansy HA, Royston TJ, Balk RA, Sandler RH (2002a) Pneumothorax detection using pulmonary acoustic transmission measurements. Med Biol Eng Comput 40(5):520–525

Mansy HA, Royston TJ, Balk RA, Sandler RH (2002b) Pneumothorax detection using computerised analysis of breath sounds. Med Biol Eng Comput 40 (5):526–532

Moore FO, Goslar PW, Coimbra R, Velmahos G, Brown CV, Coopwood TB Jr, Lottenberg L, Phelan HA, Bruns BR, Sherck JP, Norwood SH, Barnes SL, Matthews MR, Hoff WS, de Moya MA, Bansal V, Hu CK, Karmy-Jones RC, Vinces F, Pembaur K, Notrica DM, Haan JM (2011) Blunt traumatic occult pneumothorax: is observation safe? – results of a prospective, AAST multicenter study. J Trauma 70 (5):1019–1025

Myers JW, Neighbors M, Tannehill-Jones R (2002) Principles of pathophysiology and emergency medical care. Delmar Thomson Learning, Albany, p 121. ISBN 0-7668-2548-5

Oveland NP, Lossius HM, Wemmelund K, Stokkeland PJ, Knudsen L, Sloth E (2013) Using thoracic ultrasonography to accurately assess pneumothorax progression during positive pressure ventilation: a comparison with CT scanning. Chest 143(2):415–422

Plourde M, Emond M, Lavoie A, Guimont C, Le Sage N, Chauny JM, Bergeron E, Vanier L, Moore L, Allain-Boulé N, Fratu RF, Dufresne M (2014) Cohort study on the prevalence and risk factors for delayed pulmonary complications in adults following minor blunt thoracic trauma. Can J Emerg Med Care 16 (2):136–143

Roberts DJ, Leigh-Smith S, Faris PD, Ball CG, Robertson HL, Blackmore C, Dixon E, Kirkpatrick AW, Kortbeek JB, Stelfox HT (2014) Clinical manifestations of tension pneumothorax: protocol for a systematic review and meta-analysis. Syst Rev 3:3. doi:10.1186/2046-4053-3-3

Valero C, Javierre E, García-Aznar JM, Gómez-Benito MJ (2014) Nonlinear finite element simulations of injuries with free boundaries: application to surgical wounds. Int J Numer Methods Biomed Eng 30 (6):616–633

Velec M, Moseley JL, Eccles CL, Craig T, Sharpe MB, Dawson LA, Brock KK (2011) Effect of breathing motion on radiotherapy dose accumulation in the abdomen using deformable registration. Int J Radiat Oncol Biol Phys 80:265–272

Advs Exp. Medicine, Biology - Neuroscience and Respiration (2015) 15: 15–23
DOI 10.1007/5584_2015_146
© Springer International Publishing Switzerland 2015
Published online: 29 May 2015

Comparison of Small Bore Catheter Aspiration and Chest Tube Drainage in the Management of Spontaneous Pneumothorax

P. Korczyński, K. Górska, J. Nasiłowski, R. Chazan, and R. Krenke

Abstract

Beside standard chest tube drainage other less invasive techniques have been used in the management of patients with an acute episode of spontaneous pneumothorax. The aim of the study was to evaluate the short term effect of spontaneous pneumothorax treatment with small-bore pleural catheter and manual aspiration as compared to large-bore chest tube drainage. Patients with an episode of pneumothorax who required pleural intervention were enrolled in the study and randomly assigned to one of the treatment arms: (1) small-bore pleural catheter (8 Fr) with manual aspiration; (2) standard chest tube drainage (20–24 Fr). Success rate of the first line treatment, duration of catheter or chest tube drainage, and the need for surgical intervention were the outcome measures. The study group included 49 patients (mean age 46.9 ± 21.3 years); with 22 and 27 allocated to small bore manual aspiration and chest tube drainage groups, respectively. There were no significant differences in the baseline characteristics of patients in both therapeutic arms. First line treatment success rates were 64 % and 82 % in the manual aspiration and chest tube drainage groups, respectively; the difference was insignificant. Median time of treatment with small bore catheter was significantly shorter than conventional chest tube drainage (2.0 *vs.* 6.0 days; $p < 0.05$). Our results show that treatment of spontaneous pneumothorax with small-bore pleural catheter and manual aspiration might be similarly effective as is chest tube drainage in terms of immediate lung re-expansion.

P. Korczyński (✉), K. Górska, J. Nasiłowski, R. Chazan,
and R. Krenke
Department of Internal Medicine, Pneumonology and
Allergology, Warsaw Medical University, 1a Banacha St.,
02-097 Warsaw, Poland
e-mail: piotr.korczynski@wum.edu.pl

Keywords

Aspiration • Chest tube drainage • Manual aspiration • Pleural space • Small bore catheter • Spontaneous pneumothorax

1   Introduction

Spontaneous pneumothorax is defined as the presence of air in the pleural space which occurs without antecedent trauma. The cause of spontaneous pneumothorax is almost always an air leak from the lung. Based on the absence or the presence of underlying lung disease, spontaneous pneumothorax can be further subclassified as primary spontaneous pneumothorax (PSP) and secondary spontaneous pneumothorax (SSP), respectively. Chronic obstructive pulmonary disease (COPD), lung tumors, *Pneumocystis jiroveci* pneumonia, sarcoidosis, tuberculosis are the most common conditions associated with SSP (Noppen and De Keukeleire 2008). The overall incidence of spontaneous pneumothorax ranges between 24.0/100,000 for men and 9.8/100,000 for women, and admission rates are 16.7/100,000 and 5.8/100,000, respectively, in a study analyzing three national databases (Gupta et al. 2000). In that study, incidence of secondary spontaneous pneumothorax for males and females above 55 years of age is 32.4 and 10.9/100,000/year, respectively. Significant differences in severity of symptoms have been reported, with some patients being completely or near completely asymptomatic and some other presenting with acute respiratory and heart failure due to tension pneumothorax which requires an immediate intervention.

There are two major goals in the treatment strategy for spontaneous pneumothorax: to restore normal anatomical conditions by removing the air from the pleural space and to prevent recurrences. The first goal refers to all patients with an acute episode of pneumothorax, while the second applies only to patients with recurrent pneumothorax. The recurrence rates after the first episode of PSP is estimated between 23 and 37 % (Chambers and Scarci 2009; Schramel et al. 1997). Air removal from the pleural space and lung re-expansion is a common problem, which various specialists (pulmonologist, thoracic surgeons, and general surgeons) face on a daily basis. Available treatment options include conservative treatment with oxygen supplementation, manual aspiration, and chest tube drainage with underwater seal or suction (Huang et al. 2014).

Different approaches have been proposed to select the most effective and least invasive therapeutic option for the particular patient. In general, severity of symptoms, pneumothorax size and its type (primary or secondary) are the major determinants of the treatment recommended. The measurement of pneumothorax size can be done with various formulas based on the distance between visceral pleura and parietal pleura seen on the posteroanterior chest radiograph. Simple and clinically useful classification of pneumothorax size was presented in the British Thoracic Society guidelines. Large pneumothorax is defined as the distance between visceral pleura line and parietal pleura line at lung hilum level greater than 2 cm (MacDuff et al. 2010). When this distance is smaller than 2 cm pneumothorax is considered as small.

Manual aspiration (MA), placement of a small bore catheter or a standard chest tube are the main therapeutic options in symptomatic patients with spontaneous pneumothorax. Underwater seal or Heimlich valve can be used along with small bore pleural catheter and chest tube (Noppen et al. 2002). The use of a standard chest tube has a much longer history and more data on this treatment option are available. Small bore pleural catheters have been more widely applied to treat patients with spontaneous pneumothorax only in the last two decades (Kuo et al. 2013; Baumann et al. 2001). A similar success rate for MA and standard chest tube in first PSP episode was reported by Parlak et al. (2012). The authors of this study concluded that less pain associated with MA and lower hospitalization-related costs may favour MA method. To our knowledge, in Poland patients with spontaneous pneumothorax are almost exclusively treated with standard chest tubes. As compared to manual aspiration, chest tube drainage is associated with higher levels of pain, anxiety, and a longer hospital stay (Harvey and Prescott 1994). On the other hand, the literature data suggest similar immediate and long-term efficacy of MA and chest tube drainage, but the number of patients included in the studies seems to be still insufficient (Devanand et al. 2004). Thus, we undertook a comparative study of the efficacy and safety of MA *via* a small bore pleural catheter and chest tube drainage treatment in patients with spontaneous pneumothorax.

2 Methods

2.1 Study Design

The study was approved by Bioethics Committee of the Medical University of Warsaw, Poland. This prospective, randomized, parallel group, single centre trial was performed between January 2009 and June 2014. Adult patients with the first or second (first recurrence) episode of spontaneous pneumothorax who required pleural intervention (aspiration or chest tube insertion) were enrolled. The British Thoracic Society guidelines were used to select patients who should undergo pleural aspiration or chest tube insertion. Briefly, all patients with symptomatic PSP or SSP with intrapleural distance >2 cm were included (Henry et al. 2003). Pneumothorax size was assessed by the measurement of the intrapleural distance (between visceral pleura line and the line outlining the internal surface of the chest wall) at the apex, hilar level, and at the base of lung. The exclusion criteria were as follows: pregnancy, tension pneumothorax, recurrent pneumothorax within 1 year from the first episode, HIV infection, severe comorbidity, and traumatic pneumothorax. The concept of the study was discussed with each

patient on hospital admission and written informed consent was obtained. Then the patients were allocated to one of the treatment groups using computer-generated random numbers.

The first group was treated with small bore pleural catheter and manual aspiration (manual aspiration, the MA group). The second group was managed with conventional chest tube drainage (chest tube drainage, the CTD group).

Small bore catheter insertion and manual aspiration were performed as follows: (1) patient was in semi-supine position; (2) after skin disinfection and field preparation, local anaesthesia with 2 % lidocaine was applied in the second or third intercostal space at the midclavicular line; (3) a lead needle was inserted to the pleural cavity; (4) after the needle had entered the pleural space, it was directed apically and a small, 8 Fr, bore pleural catheter (Pleurocath; Prodimed, Neuilly-en-Thelle, France) was advanced into the pleural space 5–10 cm deep. The catheter was fixed to the skin and connected *via* a three-way valve to a 50-ml syringe. Then, air was manually aspirated. The procedure of aspiration was terminated when resistance was encountered and air was no longer aspirated or volume of withdrawn air exceeded 2,000 ml and there was still pleural air which could be aspirated. In the patients in whom the manual aspiration resulted in emptying the air from the pleural cavity the catheter was clamped. In those in whom pleural air was still aspirated despite the initial removal of 2,000 ml a one-way Heimlich valve was connected to the catheter. In both groups, chest X-ray was performed after 4 h with the catheter in place. If lung expansion was complete (or only a small rim of apical air was demonstrated) and there were no signs of air leak, the chest radiograph was repeated the next day and if there was no pneumothorax the catheter was removed. If no lung expansion was found in either of the chest radiographs or air leakage was observed, the Heimlich valve was used. When complete or near complete lung re-expansion was achieved and the air leak had stopped, the catheter was removed. If lung re-expansion was not achieved within 3–5 days of treatment, clinical status deteriorated, or pneumothorax size increased, the standard chest drainage was executed. If the air leak persisted up to day 7 of treatment, patients were referred for surgical intervention (video assisted thoracic surgery).

Chest tube drainage was performed using standard 20–24 Fr tubes (Portex, Smiths Medical, Dublin, OH). The procedure was done under local anaesthesia while maintaining sterile condition. The chest tube was inserted in the fourth or fifth intercostal space, between anterior and midaxillary line, and directed to the apex. The tube was fixed to the skin and connected to a water seal system. The position of the chest tube and lung re-expansion were assessed in the chest radiograph performed 4 h after tube insertion. Treatment was continued until complete or near complete lung re-expansion was achieved and air leak had stopped. The criteria for tube removal were no air leak or confirmation of lung re-expansion in two chest radiographs performed at the interval of 24 h. When no complete lung re-expansion was achieved during 7 days of treatment which chest drainage or air leak was still present, patients were referred for surgical treatment.

In both study groups, patients were discharged the next day after successful treatment, consisting of lung re-expansion, with the catheter or chest tube removed.

2.2   Outcome Measures

The primary outcome measures were defined as (1) success rate after first line treatment; (2) duration of catheter or chest tube drainage; and (3) need for surgical intervention. The success of the first line treatment in the MA group was defined as complete or nearly complete and persistent lung re-expansion after manual aspiration and the absence of air leak and removal of catheter within 5 days. Successful treatment with chest tube drainage (CTD group) was defined as complete lung re-expansion, absence of air leakage, and chest tube removal within 7 days from chest tube placement. The success rates were calculated per patient and per pleural therapeutic procedure.

The secondary outcome measures were defined as (1) success rate of second line

treatment; (2) procedure safety, and (3) duration of hospitalization. The successful second line treatment in MA group was understood as lung re-expansion and air leak cessation with subsequent chest tube removal in those patients who required chest tube insertion due to a failure of the first line treatment (small bore catheter and manual aspiration). The second line treatment could not been assessed in the CTD group as there was no such treatment in this group.

2.3   Statistical Analysis

Demographic and descriptive data are presented as median and interquartile range (IQR) or as the numbers and percentages. The differences between variables in two treatment arms were tested using the non-parametric Mann-Whitney U-test. Categorical variables were compared using Chi^2 test. A p-value <0.05 defined statistical significance. Data analysis was performed using Statistica 10.0 (StatSoft Inc., Tulsa, OK).

3   Results

Forty nine patients with spontaneous pneumothorax were enrolled and randomized to treatment arms. There were no differences between MA and CTD groups in terms of the patient age, gender, smoking history, and size of pneumothorax. Detailed patients' characteristics are presented in Table 1.

Twenty two procedures of small catheter insertion and manual aspiration were performed in the MA group. In eight patients from this group, chest tube insertion was necessary. Twenty seven chest tube insertions were done as the first line treatment in CTD group. Table 2 and Fig. 1 show the results of the study in terms of the primary and secondary outcomes in both treatment arms. The success rates after the first line treatment were somewhat higher in the CTD than in MA group, but the difference between the two groups was insignificant (81.5 vs. 63.6 %, respectively).

The median treatment duration with small bore catheter was significantly shorter than

Table 1 Patients and pneumothoraces characteristics in both treatment groups

	Manual aspiration group	Chest tube drainage group	Difference
Number of patients	22	27	
Sex, M/F; (%M)	17/5 (77.3 %)	16/11 (59.3 %)	NS
Age, median (IQR) (years)	43.5 (27.0–68.0)	43.0 (29.0–65.0)	NS
Pneumothorax side, right/left; (% right)	11/11 (50.0 %)	19/8 (70.4 %)	NS
First SP episode, n (%)	13 (59.1 %)	20 (74.1 %)	NS
Second SP episode, n (%)	9 (40.9 %)	7 (25.9 %)	NS
Apical intrapleural distance, median (IQR) (mm)	32.0 (30.0–42.7)	44.2 (31.1–82.2)	NS
Hilar intrapleural distance, median (IQR) (mm)	18.0 (10.0–25.0)	30.1 (11.4–40.0)	NS
Intrapleural distance at lung base, median (IQR) (mm)	15.6 (10.5–36.0)	31.6 (17.4–50.0)	$p < 0.05$
Smoking history, Yes/No (%Yes)	13/9 (59.1 %)	15/12 (51.7 %)	NS

M/F male/female, *IQR* interquartile range, *SP* spontaneous pneumothorax

Table 2 The results of the study (primary and secondary outcome measures)

Outcomes	Manual aspiration group (n = 22)	Chest tube drainage group (n = 27)	Difference
First line treatment success rate, n (% of all in the treatment arm)	14 (63.6 %)	22 (81.5 %)	NS
Secondary line treatment success rate – secondary chest tube drainage	6 (27.3 %)	–	NA
Surgical intervention, n (% of all patients in treatment arm)	2 (9.1 %)	5 (18.5 %)	NS
Medical (non-surgical) treatment success rate, n (% of all patients in treatment arm)	20 (90.9 %)	22 (81.5 %)	NS
Drainage duration, days (IQR)	2.0 (1.0–5.0)	6.0 (2.0–8.0)	$p < 0.05$
Hospital stay, days (IQR)	4.0 (3.0–8.0)	7.0 (5.0–12.0)	$p < 0.05$

NS non-significant, *NA* not applicable, *IQR* interquartile range

conventional chest tube drainage, and this difference was associated with a shorter hospital stay. Failure of the first line treatment was noted in eight patients in the MA group. Six of these patients were further successfully treated with standard chest tube. Thus, in 20 of the 22 patients in the MA group treatment goals were achieved with medical treatment and only 2 (9 %) patients required surgical intervention. In the CTD group, first line therapy was unsuccessful in five patients.

There were eight patients with the first episode of SSP. In all these patients, first line therapy was found successful (two patients in MA and six patients in CTD group). The first line therapy was less successful in patients with the second episode of SSP (75 % and 66 % success rate in MA and CTD groups, respectively). In both groups treatment was well

tolerated and no severe complications were observed.

4 Discussion

The present study compared treatment of spontaneous pneumothorax with manual aspiration and chest tube drainage. Although there are studies demonstrating that the small bore pleural catheter with manual aspiration may have some advantages over the chest tube drainage (Noppen et al. 2002), the former method is only incidentally used in Poland. Thus, the present study seemed important for the eventual encouragement for a wider use of small bore pleural catheters with manual aspiration in the management of patients with spontaneous

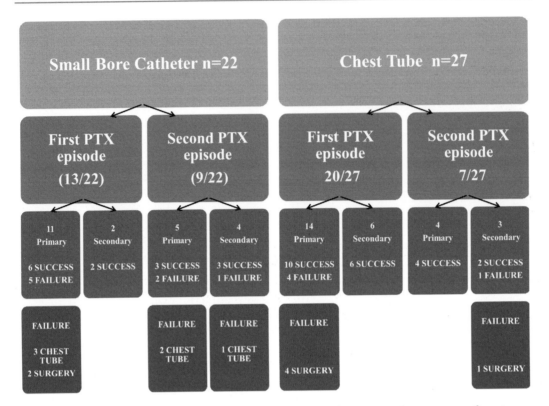

Fig. 1 The flowchart presenting the results of treatment in the context of treatment regimen, category of spontaneous pneumothorax (primary or secondary) and pneumothorax episode (first or second). *PTX* pneumothorax

pneumothorax. We found no statistical difference between small bore manual aspiration and chest tube drainage in terms of first line treatment efficacy. The goal of treatment was achieved in about 64 % of patients managed with small bore aspiration and 82 % of patients treated with chest tube drainage. Although the difference was statistically insignificant, the success rate was somehow shifted toward the latter method, which might be of clinical relevance. Also, we cannot exclude that a larger number of participants in both treatment arms would increase the statistical power of the difference to the advantage of chest tube drainage. The present results are in line with a study of Ayed et al. (2006) who have found comparable treatment efficacy in patients managed with small bore manual aspiration (62 %) and chest tube drainage (68 %).

It should be noted that comparison of the results in our both treatment arms may raise some doubts. While the percentage of patients who were successfully treated with small bore manual aspiration was slightly lower than that with chest tube drainage (64 *vs.* 82 %), the percentage of patients referred to surgery was almost 10 % lower in case of manual aspiration than that in chest tube drainage. This was because standard chest tube drainage was applied as a second line treatment in those patients in whom treatment with small bore pleural catheter with manual aspiration had failed. Thus, referral for surgical intervention in small bore manual aspiration resulted from the unsatisfactory effect of the first and second line treatment; i.e., small bore manual aspiration and chest tube drainage, while in the other group surgical intervention resulted from the unsatisfactory effect of just chest tube drainage. In this context, a 64 % success rate in patients treated with small bore catheter with manual aspiration might be regarded as a true benefit that reflects the number of patients in whom treatment with chest tube drainage could have been avoided.

Our study involved only patients with spontaneous pneumothorax. In other studies, which evaluated various aspects of pneumothorax management with small bore pleural catheter, patients with traumatic pneumothorax were also included. A large group of 212 patients with pneumothorax was treated by Contou et al. (2012). This group included 117 (55 %) patients with primary spontaneous pneumothorax, 28 (13 %) of patients with secondary pneumothorax associated with COPD, 19 (9 %) patients had traumatic pneumothorax, and 48 (23 %) had iatrogenic pneumothorax. The authors reported a similar success rate in two treatment arms which used small bore central venous catheter (5 F) and chest tube (14–20 F) drainage. Failure rates in both groups were 18 and 21 %, respectively. Interestingly, the highest success rate was achieved in patients with iatrogenic and traumatic pneumothorax. Ninety six percent of patients with iatrogenic and 67 % of patients with traumatic pneumothorax were successfully treated with small bore catheter. The respective numbers reported for chest tube drainage were 100 and 92 %.

Cho and Lee (2010) have reported a significant difference in treatment efficacy in patients with PSP and SSP. Small bore 7 F catheter was used in this study. Eighty three percent of patients with PSP were successfully treated as compared to only 47.4 % of patients with SSP The first line treatment with small bore manual aspiration resulted in about 56 % success rate (9 of the 16 patients) in patients with PSP and a 83 % success rate in patients with SSP (5 of the 6 patients). Those results are different from the present ones. However, the present results should be interpreted with caution as the number of patients with PSP and SSP in small bore manual aspiration group was several-fold smaller than that in the study above cited. We also analyzed the results of treatment in the context of the first or second episode of pneumothorax and found no appreciable differences depending on pneumothorax recurrence. That concerns the patients treated with both small bore manual aspiration

(62 *vs.* 66 %, respectively) and with chest tube drainage (80 *vs.* 86 %, respectively). The relationship between the treatment outcome and pneumothorax episode (first or recurrent) has also been a subject of research done by Ho et al. (2011). The authors compared mini-chest tube with needle aspiration in outpatient management of PSP. They found no association between the efficacy of treatment and the first or recurrent episode of pneumothorax. Nor could they substantiate a relationship between treatment outcome and other covariates such as age, sex, smoking history, and size of pneumothorax.

In the present study, simple aspiration resulted in a shorter drainage time than in case of chest tube drainage and in a decrease of hospitalization duration. These findings are compatible with the studies demonstrating that small bore catheter placement with manual aspiration can significantly reduce the length of hospital stay in patients with pneumothorax (Contou et al. 2012). Moreover, simple aspiration is a non-invasive method that can be safely performed on the outpatient basis (Lai and Tee 2012). Noppen et al. (2002) have reported that only 52 % of patients with PSP treated with manual aspiration required hospitalization. In contrast 100 % of patients treated with chest tube drainage had to be hospitalized.

We did not observe any serious complications neither in MA nor in CTD treatment arm. This is consistent with other reports. Some minor complications, such as subcutaneous emphysema, and catheter kinking or dislodgement, have been observed in a study of Cho and Lee (2010), without any mortalities or major complications, such as bleeding, empyema/infection, inadequate chest tube positioning, or pulmonary laceration. The present results also are consistent with the safety profile reported by Benton and Benfield (2009), which clearly favors small caliber drainage tube. In another study, the only complication noted in one of the 24 patients treated with small-bore drain (12 F) was subcutaneous emphysema, whereas the large-bore drain (20–24 F) had a higher complication rate (32 %), including the drain site and intrapleural infections. The authors admitted, however, that small bore tube was more prone to displacement (Ayed et al. 2006).

A recurrence rate is in an important issue when treating patients with spontaneous pneumothorax. However, this aspect was beyond of the scope of our study. We had no sufficient follow-up data enabling reliable assessment of the recurrence rate in both treatment arms. However, it is a common belief that the methods used in the treatment of acute spontaneous pneumothorax episode have null or very little influence on the recurrence of spontaneous pneumothorax (Zehtabchi and Rios 2008). This is also a conclusion from a study of Ayed et al. (2006), which included 137 patients with the first episode of PSP. Patients were randomly assigned to treatment with simple aspiration or tube thoracostomy. The recurrence rate after 1 and 2 years were 22 % and 31 %, respectively, in patients with simple aspiration, and 24 % and 25 % in those treated with tube thoracostomy.

The present study has some limitations. The number of patients in both therapeutic arms was relatively small. As a minimal clinical difference in the primary and secondary outcome measures (the success rate, drainage duration, and hospital stay) was difficult to define, the number of patients in each therapeutic arm could not be precisely predetermined. Thus, statistical differences between the results observed in two treatment arms should be interpreted with caution. On the other hand, the total number patients in our study was only slightly lower than the 56 patients in a study of Parlak et al. (2012) or 60 patients in that of Noppen et al. (2002). Comparison of the present results might be confounded by the fact that treatments applied in small bore manual aspiration and chest tube drainage groups were not equivalent. Two lines of treatment were undertaken in the former and just one in the latter group before referring patient for surgical treatment. However, it is common clinical practice that patients in whom manual aspiration fails are almost always subsequently treated with chest tube drainage. On the other hand, only in very small and carefully selected patients reinsertion of a new chest tube might be considered as a second line therapeutic option when earlier unsuccessfully treated with chest tube drainage. The efficacy of treatment with manual aspiration in our study could have been affected by a lower experience of participating physicians with the small-bore catheter procedure, a relatively new technique, as opposed to standard chest drainage mostly used in the management of pneumothorax. Finally, some aspects, such as the duration of the procedure, the involvement of medical staff, the treatment costs, and the effect on the recurrence rate, were left out from the assessment in the present study.

In conclusion, the current study shows that small-bore pleural catheter with manual aspiration is an effective treatment method for patients with the first or recurrent episode of spontaneous pneumothorax. This management strategy may significantly reduce the number of patients treated with large bore chest tube drainage and may shorten drainage duration.

Conflicts of Interest The authors declare no conflicts of interest in relation to this article.

References

Ayed AK, Chandrasekaran C, Sukumar M (2006) Aspiration versus tube drainage in primary spontaneous pneumothorax: a randomised study. Eur Respir J 27:477–482

Baumann M, Strange C, Heffner J, Light R, Kirby T, Klein J, Luketich J, Panacek E, Sahn S (2001) Management of spontaneous pneumothorax: an American College of Chest Physicians Delphi consensus statement. Chest 119:590–602

Benton I, Benfield G (2009) Comparison of a large and small-calibre tube drain for managing spontaneous pneumothoraces. Respir Med 103:1436–1440

Chambers A, Scarci M (2009) In patients with first-episode primary spontaneous pneumothorax is video-assisted thoracoscopic surgery superior to tube thoracostomy alone in terms of time to resolution of pneumothorax and incidence of recurrence? Interact Cardiovasc Thorac Surg 9:1003–1008

Cho S, Lee E (2010) Management of primary and secondary pneumothorax using a small-bore thoracic catheter. Interact Cardiovasc Thorac Surg 11:146–149

Contou D, Razazi K, Katsahian S, Maitre B, Mekontso-Dessap A, Brun-Buisson C, Thille A (2012) Small-bore catheter versus chest tube drainage for pneumothorax. Am J Emerg Med 30:1407–1413

Devanand A, Koh M, Ong T, Low S, Phua G, Tan K, Philip E, Samuel M (2004) Simple aspiration versus chest-tube insertion in the management of primary spontaneous pneumothorax: a systematic review. Respir Med 98:579–590

Gupta D, Hansell A, Nichols T, Duong T, Ayres JG, Strachan D (2000) Epidemiology of pneumothorax in England. Thorax 55:666–671

Harvey J, Prescott RJ (1994) Simple aspiration versus intercostal tube drainage for spontaneous pneumothorax in patients with normal lungs. British Thoracic Society Research Committee. Br Med J 309:1338–1339

Henry M, Arnold T, Harvey J (2003) BTS guidelines for the management of spontaneous pneumothorax. Thorax 58(Suppl 2):ii39–ii52

Ho KK, Ong MEH, Koh MS, Wong E, Raghuram J (2011) A randomized controlled trial comparing minichest tube and needle aspiration in outpatient management of primary spontaneous pneumothorax. Am J Emerg Med 29:1152–1157

Huang Y, Huang H, Li Q, Browning RF, Parrish S, Turner JF, Zarogoulidis K, Kougioumtzi I, Dryllis G, Kioumis I, Pitsiou G, Machairiotis N, Katsikogiannis N, Courcoutsakis N, Madesis A, Diplaris K, Karaiskos T, Zarogoulidis P (2014) Approach of the treatment for pneumothorax. J Thorac Dis 6:S416–S420

Kuo H, Lin Y, Huang C, Chien S, Lin I, Lo M, Liang C (2013) Small-bore pigtail catheters for the treatment of primary spontaneous pneumothorax in young adolescents. Emerg Med J 30:e17

Lai S, Tee A (2012) Outpatient treatment of primary spontaneous pneumothorax using a small-bore chest drain with a Heimlich valve: the experience of a Singapore emergency department. Eur J Emerg Med 19:400–404

MacDuff A, Arnold A, Harvey J (2010) Management of spontaneous pneumothorax. British Thoracic Society pleural disease guideline 2010. Thorax 65:Ii18–Ii31

Noppen M, De Keukeleire T (2008) Pneumothorax. Respiration 76:121–127

Noppen M, Alexander P, Driesen P, Slabbynck H, Verstraeten A (2002) Manual aspiration versus chest tube drainage in first episodes of primary spontaneous pneumothorax: a multicenter, prospective, randomized pilot study. Am J Respir Crit Care Med 165:1240–1244

Parlak M, Uil SM, van den Berg JWK (2012) A prospective, randomised trial of pneumothorax therapy: manual aspiration versus conventional chest tube drainage. Respir Med 106:1600–1605

Schramel FM, Postmus PE, Vanderschueren RG (1997) Current aspects of spontaneous pneumothorax. Eur Respir J 10:1372–1379

Zehtabchi S, Rios CL (2008) Management of emergency department patients with primary spontaneous pneumothorax: needle aspiration or tube thoracostomy? Ann Emerg Med 51:91–100

Advs Exp. Medicine, Biology - Neuroscience and Respiration (2015) 15: 25–34
DOI 10.1007/5584_2015_149
© Springer International Publishing Switzerland 2015
Published online: 29 May 2015

Brown Adipose Tissue and Browning Agents: Irisin and FGF21 in the Development of Obesity in Children and Adolescents

B. Pyrżak, U. Demkow, and A.M. Kucharska

Abstract

In the pediatric population, especially in early infancy, the activity of brown adipose tissue (BAT) is the highest. Further in life BAT is more active in individuals with a lower body mass index and one can expect that BAT is protective against childhood obesity. The development of BAT throughout the whole life can be regulated by genetic, endocrine, and environmental factors. Three distinct adipose depots have been identified: white, brown, and beige adipocytes. The process by which BAT can become beige is still unclear and is an area of intensive research. The "browning agents" increase energy expenditure through the production of heat. Numerous factors known as "browning agents" have currently been described. In humans, recent studies justify a notion of a role of novel myokines: irisin and fibroblast growth factor 21 (FGF21) in the metabolism and development of obesity. This review describes a possible role of irisin and FGF21 in the pathogenesis of obesity in children.

Keywords

Beige adipose tissue • Irisin • FGF21 • Obesity • Children • White adipose tissue

B. Pyrżak (✉) and A.M. Kucharska
Department of Pediatrics and Endocrinology, Medical University of Warsaw, 24 Marszalkowska St., 00-576 Warsaw, Poland
e-mail: beata.pyrzak@wum.edu.pl

U. Demkow
Department of Laboratory Diagnostics and Clinical Immunology of the Developmental Age, Medical University of Warsaw, Warsaw, Poland

1 Introduction

The prevalence of obesity in the pediatric population has increased over the past few decades. There is a growing body of evidence that obesity in childhood predicts the risk of cardiometabolic diseases and diabetes mellitus type 2 later in life. Adipose tissue has a central role in the regulation of energy balance and homoeostasis. There are two main types of adipose tissue: white adipose tissue (WAT) and brown adipose tissue (BAT).

WAT serves as the primary site of energy, BAT is highly vascularized, rich in mitochondria and highly metabolically active. In contrast to the energy storing function of WAT, brown adipose tissue present in small mammals and newborns has long been considered as protective against a body temperature drop. That process requires increased energy consumption due to thermogenesis taking place by uncoupling the electron transport chain from ATP synthesis *via* uncoupling protein-1 (UCP-1) (Van Marken Lichtenbelt et al. 2009; Cannon and Nedergaard 2004).

BAT is most abundant in infancy, a period of increased susceptibility to hypothermia due to low skeletal muscle content and high body surface area in relation to volume (Tews and Wabitsch 2011). Functional BAT persists beyond infancy, distributed primarily in the supraclavicular, neck, paravertebral, and suprarenal areas of the body (Virtanen et al. 2009). The presence and volume of functional BAT increases in the period of puberty and decreases later in life, inversely correlating with body mass index and being more active in females than in males (Drubach et al. 2011).

Infancy is a period of transition when both the pre- and post-natal environments contributes to body composition. A recent meta-analysis has found that infants with high birth weight and with intrauterine growth restriction had increased risk for the development of obesity later in life (Yu et al. 2011), which constitutes a strong risk factor for the development of metabolic syndrome, insulin resistance, and cardiovascular disease (Ornoy 2011; Barker et al. 1993). It is also known that children of diabetic mothers can be heavier and may possess more adipose tissue at birth. It is unclear if BAT contributes to the development of obesity in these children.

2   Stages of BAT Development

The origin and development of WAT and BAT in pre and postnatal life in humans is still a matter of debate. The characteristics of primary stages of BAT development in small animals has been described by Symonds (2013). Adipocytes are generated in mid-gestation, have multilocular appearance, but do not express thermogenin (called uncoupling protein-1 or UCP-1) (Clarke et al. 1997a). During maturation the fat tissue contains a mixture of unilocular (rich in lipids) and multilocular (abundant in mitochondria and expressing UCP-1) cells (Gemmell et al. 1972; Pope et al. 2014). It is unknown how the proportion between BAT and a mixture of beige and white adipocytes is regulated, although it is clear that at least four distinct stages of adipose tissue development exist in early life. By mid-gestation when adipose tissue becomes detectable, it has a dense cellular structure. Close to term, as the depot increases in size, cells with the appearance of both white and brown adipocytes surrounded the larger, lipid filled (white) cells are visible (Gemmell and Alexander 1978). In the neonatal period, a pronounced reduction in the number of white adipocytes occurs coincidently with maximal UCP1 abundance. Finally, brown adipocytes disappear gradually throughout the postnatal period, with only white adipocytes being discernable by 1 month of age (Clarke et al. 1997a; Pope et al. 2014).

Symonds (2013) has described regulatory factors in small animals including hormones, adipokines, and selected genes which are involved in the activity of adipose tissue in the following phases: early gestation, at birth, and after birth. In the proliferative phase of early mid-gestation, the expression of cellular proliferation marker Ki-67 and genes involved in the cell development, such as the homeobox (HOX) and bone morphogenic proteins (BMP): HOXA1, HOXC9, BMP4, and BMP7, is observed (Tseng et al. 2008; Scholzen and Gerdes 2000).

2.1   Preparatory Phase for Thermogenesis Immediately After Birth

The substantial growth of the perirenal-abdominal depot up to term is mediated by CCAAT-enhancer-binding protein β (C/EBPβ), which forms a transcriptional complex critical for adipogenesis (Clarke et al. 1997a; Kajimura et al. 2009). The second stage of adipogenesis is characterized by significant abundance of UCP-1 (Pope et al. 2014). This process is primarily regulated by endocrine stimulatory factors which act to maximize the amount and thermogenic potential of UCP-1 (Symonds et al. 1995; Bird et al. 1996). The gene coding of the long form of the prolactin receptor (PRLR) is highly expressed prior to birth (Pope et al. 2014), which is in accordance with its crucial role in thermogenesis demonstrated in both small and large mammals (Pearce et al. 2005; Viengchareun et al. 2008). Furthermore, the peak in deiodinase iodothyronine type II (DIO2) activity provides triiodothyronine, which can activate several genes essential for UCP-1 function, including peroxisome proliferator-activated receptor-gamma (PPARγ) coactivator (PGC-lα) and β₃-adrenergic receptor (β3AR) (Bassett and Symonds 1998; Uldry et al. 2006; Ribeiro et al. 2010; Hall et al. 2010). These proteins, in conjunction with PPARγ, facilitate BAT thermogenesis (Seale 2010). The depot size peaks just prior to birth and the gene expression of Ki-67 declines to the basal level indicating that the replicative period is finished and cells are committed to terminal differentiation (Clarke et al. 1997a; Gregoire et al. 1998).

2.2 Birth – Shift Toward Non-shivering Thermogenesis

Prior to birth, fetal BAT contains a large number of lipid filled cells which become rapidly depleted postnatal period, as heat production is maximized (Clarke et al. 1997b) and abundance of UCP-1 peaks. This process is coincident with significant falls in the expression of white adipose tissue marker genes, such as adiponectin, leptin, and corepressor receptor-interacting protein 140 (RIP140) (Pope et al. 2014). This phase is also characterized by an increase in C/EBPβ and BMP4, which could be indicative of a change over to the white adipocyte-like phenotype.

2.3 Loss of UCP1 and Accumulation of Lipid in the Perirenal-Abdominal Depot

The final phase during early postnatal life coincides with the loss of UCP-1 (Clarke et al. 1997b) and the appearance of white adipose tissue. Adipocyte size increases and the peak expression for a number of genes representative for mature adipocytes occurs, including genes encoding: adiponectin, leptin, receptor-interacting protein 140 (RIP140), PPARγ, BMP7, HOXC9, and glucocorticoid receptor 2. The loss of UCP-1 is accompanied by a declining expression of genes primarily associated with BAT, such as PRLR, or peroxisome proliferator-activated receptor gamma coactivator 1-alpha (PGC1α); the process facilitated by a concomitant rise in twist-related protein TWIST1, as it is a transcriptional repressor that interacts directly with PGC-1α to suppress both thermogenic and mitochondrial transcription factors (Pan et al. 2009). The expression of a number of other genes indicative of a change in adipocyte cell number also peaks at one month of age; the primary changes of note are increases in cell size together with the loss of a BAT phenotype (Pope et al. 2014). Continuing differentiation of white adipocytes is possible as indicated by the rise in the expression of BMP4 (Pope et al. 2014) and BMP7 (Schulz and Tseng 2009). Postnatal adipose tissue growth is also regulated by glucocorticoid receptor 2 which promotes the UCP-1 action and further gene expression, along with the increase of white fat mass throughout postnatal and juvenile life (Mostyn et al. 2003; Gnanalingham et al. 2005). It is possible that this depot comprises beige adipocytes that have the capacity to respond to a challenge (Pope et al. 2014).

After birth several potent stimulators rapidly initiate nonshivering thermogenesis (Symonds et al. 2003). A large number of endocrine factors have a potential to activate BAT in newborns including: catecholamines (Symonds et al. 2000), thyroid hormones (Heasman et al. 2000), cortisol (Clarke et al. 1998), leptin

(Mostyn et al. 2002, 2014), and prolactin (Pearce et al. 2005). The release of these factors declines over the first few weeks of life thermogenesis from shivering to nonshivering mode as a dominant response to cold exposure (Symonds et al. 1989; Bispham et al. 2002). Other organs and tissues can also promote BAT function. The postnatal maturation of BAT mediated by the release of fibroblast growth factor (FGF21) is influenced by the onset of feeding and initiation of hepatic function (Hondares et al. 2010). FGF21 activates BAT in response to cold exposure. Further, FGF21 stimulates the accumulation of brown-like cells in WAT after cold exposure and is an upstream effector of adiponectin, which controls the general expenditure of energy (Ohta and Itoh 2014). A cross link between functional BAT and muscle volume has been recently suggested in children and adolescents. Pediatric patients with BAT visible in positron emission tomography-computed tomography images have a significantly greater muscle volume than patients without detectable BAT (Gilsanz et al. 2011).

Recent experiments have highlighted that BAT and muscles share important features. Brown fat cell precursors express a wide array of muscle-related genes (Timmons et al. 2007). The ontogenic relationship between BAT and skeletal muscles may explain why brown fat cells are specialized for lipid catabolism rather than storage. Further, BAT and skeletal muscles are both highly responsive to sympathetic nerve activity and have the capacity for adaptive thermogenesis. An *in vivo* fat mapping demonstrates that brown but not white fat cells, arise from precursors that express the *myf5* gene, a gene previously thought to be expressed only in the myogenic lineage. A transcriptional regulator, PR domain containing 16 PRDM16, controls a bidirectional switch between skeletal myoblasts and brown fat cells. Loss of PRDM16 from brown fat precursors causes loss of brown fat characteristics and promotes muscle differentiation. Conversely, ectopic expression of PRDM16 on myoblasts induces their differentiation into brown fat cells. PRDM16 stimulates brown adipogenesis by binding to PPARγ and activating

its transcriptional function. Finally, PRDM16-deficient brown fat displays an abnormal morphology, reduced thermogenic gene expression, and an elevated expression of muscle-specific genes (Seale et al. 2010).

3 Transdifferentiation Between White and Brown Adipose Tissue

In adulthood, depots of WAT, when subjected to certain stimuli, could take on a BAT phenotype called the browning of WAT. The brown adipocytes – beige adipocytes have low thermogenesis activity and a small number of mitochondria at basal state. Once activated, they possess many biochemical and morphological features of BAT, such as the presence of multilocular lipid droplets and multiple mitochondria. The inducible expression of UCP-1 and other genes related to mitochondrial biogenesis in response to appropriate external stimulation results in a cell type capable for adaptive thermogenesis and energy expenditure similar to classical BAT. Indeed, browning of WAT has been shown to have anti-obesity and antidiabetic effects in rodents (Wu et al. 2013).

Some authors argue that beige adipocytes could be derived from WAT in response to appropriate stimuli, others suggest that beige adipocytes are a new type of adipocytes derived from progenitors that are distinct from those for WAT and BAT. Wu et al. (2012) have shown that a subset of precursor cells within murine subcutaneous adipose tissue assume a gene expression pattern distinct from both white and classical brown adipocytes in response to forskolin treatment. Further, these cells could give rise to beige cells (Wu et al. 2012). Studies using positron emission tomography-computed tomography have revealed a high prevalence of metabolically active regions around the supraclavicular areas in adults. Biopsies from these regions are enriched with UCP-1-positive cells (Taittonen et al. 2009; Van Marken Lichtenbelt et al. 2009; Zingaretti et al. 2009). BAT activity negatively correlates with body mass index and body fat (Cypess et al. 2009; Saito et al. 2009; Van Marken Lichtenbelt et al. 2009). Markers that are selective for brown or beige adipocytes identified in rodents have provided a tool to investigate the molecular signature of UCP-1-positive cells in human adults (Wu et al. 2013). These studies highlight that humans, similarly to rodents, possess both classical brown and beige fat.

Lo and Sun (2013), on the basis of mouse studies, proposed a list of several representative transcriptional factors and co-regulators that affect the browning of WAT. Additionally, several hormones and non-transcriptional regulators are involved in this process. Brown adipogenesis is an area of limited understanding. It is known, however, that irisin and FGF21 are notably influential in WAT browning. These regulators seem also the best candidates to be studied in children and adolescents.

4   Irisin (FNDC5) and Its Role in Obesity in Humans

A myokine, irisin, has been described by Boström et al. (2012). Irisin is released upon cleavage of the plasma membrane protein fibronectin type III domain containing protein 5 (FNDC5). The FNDC5 gene encodes a prohormone, a single-pass type I membrane protein that is upregulated by muscular exercise and undergoes post-translational processing to generate irisin. The protein sequence includes a signal peptide, two fibronectin type III (FNIII) domains, and a C-terminal hydrophobic domain that is probably anchored in the cell membrane. The production of irisin shares several typical features with other hormones and hormone-like polypeptides, such as epidermal growth factor EGF or transforming growth factor TGFα, released from transmembrane precursors. After the N-terminal signal peptide is removed, the peptide is proteolytically cleaved from the C-terminal moiety, glycosylated, and released as a hormone of 112 amino acids, which comprises most of the FNIII repeat region. The sequence of irisin is highly conserved in mammals; the human and murine sequences are identical. Tissue distribution analysis has confirmed that FNDC5 is predominantly expressed in human muscles. FNDC5 is likely to be expressed not only in skeletal muscles but also in cardiac and smooth muscles. A relatively high expression of this gene has also been found in the pericardium, intracranial artery, and rectum. A low expression of FNDC5 mRNA is observed in major organs and tissues, such as kidney, liver, lung, and fat. Irisin is present in higher concentration in the subcutaneous white adipose tissue (Boström et al. 2012).

The beneficial effects of exercise, reduction of diet-induced obesity, and a decrease of insulin resistance in mice are attributed to irisin (Boström et al. 2012). Its effects on the browning were demonstrated in the subcutaneous WAT in muscle-specific PGC-1α transgenic mice. The effects of FNDC5 are mediated, in part, by the transcriptional factor PPARα as its antagonist significantly decrease production of UCP-1. After adenoviral delivery of FNDC5 to the liver, the plasma concentration of irisin increased resulting in further browning of subcutaneous WAT, which accounts for the protection against diet-induced obesity and insulin resistance (Boström et al. 2012). A large genetic study by Raschke et al. (2013) provided evidence for the existence of a mutation in the conserved start codon of the human FNDC5 gene, switching the ATG start codon to the alternative codon ATA. HEK293 cells transfected with the human FNDC5 construct with ATA as start codon resulted in only 1 % full-length protein compared with the human FNDC5 with ATG. That study also shows that the FNDC5 mRNA expression in muscle biopsies from humans is not changed by endurance or strength of exercise. Preadipocytes isolated from human subcutaneous adipose tissue differentiated to bright human adipocytes when incubated with bone morphogenetic protein BMP7, but neither recombinant FNDC5 nor irisin were effective. The authors suggest that some function of FNDC5 and irisin observed in mice might be lost in humans.

Several studies have reported beneficial effects of exercise in humans, such as weight loss and improved glucose tolerance, while others have questioned these findings. Staiger et al. (2013) in a group of 1976 adults described FNDC5 tagging of SNPs variants (rs16835198 and rs726344), accounting for improvement in oral glucose tolerance tests. This finding provides evidence that the FNDC5 gene encoding irisin determines insulin sensitivity in humans. Conversely, Norheim et al. (2014) in a group of 26 inactive men have observed a decrease in circulating irisin in response to 12 weeks of training; however just after acute exercise irisin was transiently increased. The plasma concentration of irisin was higher in pre-diabetic subjects compared with controls. The effect of 12 weeks of training on the expression of selected browning genes in subcutaneous adipose tissue was insignificant. UCP1 mRNA did not correlate with FNDC5 expression in the subcutaneous adipose tissue, skeletal muscle, or irisin in the plasma. Further, the authors did not observe any effect of long-term training on the level of circulating irisin, and a little or no effect on the browning of subcutaneous WAT.

Boström et al. (2012) proposed that irisin promotes the conversion of white fat to brown fat in humans, which would make it a health promoting hormone. Timmons et al. (2012) noted that over 1,000 genes are upregulated by exercise. Of note, FNDC5 was upregulated only in highly active elderly subjects, in contrast to Boström et al.'s (2012) conclusion. Over 30 studies have attempted to assess irisin in the human plasma using different, validated methods (Huh et al. 2012). Wrann et al. (2013) have shown that the hippocampal expression of FNDC5 in mice is induced by prolonged exercise. Accordingly, peripheral delivery of FNDC5 to the liver by adenoviral vectors increases the level of circulating irisin, activates a neuroprotective gene program in the brain, including the expression of brain-derived neurotrophic factor (BDNF). Blüher et al. (2014) have analyzed irisin, adipokines (leptin, adiponectin, and resistin), C-reactive protein (CRP), and soluble tumor necrosis factor receptor II (sTNFR-II) before and after exercise in obese children. The authors show that a yearlong lifestyle intervention program is associated with an improvement in anthropometric and metabolic parameters and an elevation of irisin level. In another study, Al-Daghri et al. (2014) have examined the anthropometrics, glycemic profile, lipid, adipocytokine profile, and irisin in 120 families. The authors sought to determine the relationships with heritability of irisin in parents and children. The most significant heritable traits between mother and son included: irisin, systolic blood pressure, total cholesterol, and LDL-cholesterol. Heritable traits between mother and daughter also comprised irisin, anthropometric associations, such as waist and hip circumference, blood pressure, HDL-cholesterol, and tumor necrosis factor-alpha (TNF-α). HDL-cholesterol was the single most significant predictor of irisin level in adults, explaining 17 % of the variance, whereas in children angiotensin II was the most significant predictor of irisin level, explaining 19 % of the variance. The authors conclude that circulating irisin appears to be maternally inherited and its level can be predicted from that of HDL-cholesterol in adults and from angiotensin II level in children; both factors influenced by energy expenditure and regulation. These findings confirm a significant role of irisin in the energy-generating processes. The long term clinical implications of these findings, the molecular pathways through which this novel myokine exerts its actions, and the potential impact of exercise regimen, duration and intensity on irisin level need to be further elucidated.

5 Fibroblast Growth Factor 21 (GF21) and Its Role in Obesity in Humans

The FGF21 is a new metabolic hormone. It belongs to the FGF gene family; is dominantly, but not exclusively expressed in the liver. FGF21 is mostly induced by stress and acts through the FGF receptor 1c with β-Klotho as a cofactor in the endocrine, and in part autocrine/paracrine, manner. Hepatic FGF21 directly acts on white adipocytes to inhibit lipolysis and acts through the brain to increase the systemic glucocorticoid level and to suppress physical activity in response to starvation. Administering FGF21, both *in vivo* and *in vitro*, increases the expression of UCP1 and other brown-fat-related genes in perirenal and inguinal WAT. It also protects against dioxin toxicity. Adipocytic FGF21 induces the browning of WAT and activates brown adipocytes in response to cold. It also acts as an upstream effector of adiponectin in white adipocytes. Myocytic FGF21 protects against diet-induced obesity and insulin resistance, induces the browning of WAT, and protects against cardiac hypertrophy. In addition, FGF21 polymorphism is possibly related with metabolic diseases and the FGF21 is a biomarker of metabolic disorders. These findings indicate that FGF21 plays roles as a hepatokine, adipokine, and a myokine in the metabolism, injury protection, and diseases.

In the adult population, increased serum levels of FGF21 are described in the coronary heart disease, atherosclerosis, obesity and type 2 diabetes, mitochondrial diseases, Cushing's syndrome, and preeclampsia, and decreased levels of FGF21 are in anorexia nervosa (Itoh 2014). Bisgaard et al. (2014) have investigated the oral glucose tolerance, insulin, FGF21, adiponectin, and leptin in relation to dual energy X-ray absorptiometry (DXA) scans and pubertal staging in 179 children and adolescents in the Copenhagen Puberty Study. The findings were that girls had a significantly higher level of FGF21 compared with boys. The baseline level of FGF21 positively correlated with triglycerides, but not with the body mass index, DXA-derived fat percentage, LDL, HDL, and non-HDL cholesterol, leptin, or adiponectin. Nor was there any correlation between the baseline FGF21 level and dynamic changes in glucose and insulin regarding the glucose tolerance. The authors state that FGF21 is independent of adiposity in children and its metabolic effect seems limited to pathological conditions associated with insulin resistance. By contrast, Giannini et al. (2013) have found an increase in FGF21 in obese adolescents, which positively correlates with the hepatic fat content. Consequently, FGF21 significantly correlated with the adiposity indices, visceral fat, hepatic fat content, cytokeratin 18, and alanine aminotransferase. In subjects with steatohepatitis, FGF21 correlated with the activity score of nonalcoholic fatty liver disease, independently of body mass index, visceral fat, and insulin sensitivity.

6 Conclusions

An in depth analysis of the browning process is warranted to discern possibilities to target brown and bright adipose tissue to counteract human obesity. It is of utmost importance is to shed light on the origin of fat tissues in humans, especially in obese children. The mechanism of action of irisin, the influence of exercise on it, and a link between the fibroblast growth factor 21 and metabolic complications are the basic issues to be resolved.

Conflicts of Interest The authors declare no conflicts of interests in relation to this article.

References

Al-Daghri NM, Al-Attas OS, Alokail MS, Alkharfy KM, Yousef M, Vinodson B, Amer OE, Alnaami AM, Sabico S, Tripathi G, Piya MK, McTernan PG, Chrousos GP (2014) Maternal inheritance of circulating irisin in humans. Clin Sci (Lond) 126(12):837–844

Barker DJ, Hales CN, Fall CH, Osmond C, Phipps K, Clark PM (1993) Type 2 (non-insulin-dependent) diabetes mellitus, hypertension and hyperlipidaemia (syndrome x): relation to reduced fetal growth. Diabetologia 36:62–67

Bassett JM, Symonds ME (1998) β2-agonist ritodrine, unlike natural catecholamines, activates thermogenesis prematurely in fetal sheep. Am J Physiol 275(1): R112–R119

Bird JA, Spencer D, Mould T, Symonds ME (1996) Endocrine and metabolic adaptation following caesarean section or vaginal delivery. Arch Dis Child Fetal Neonatal Ed 74(2):F132–F134

Bisgaard A, Sørensen K, Johannsen TH, Helge JW, Andersson AM, Juul A (2014) Significant gender difference in serum levels of fibroblast growth factor 21 in Danish children and adolescents. Int J Pediatr Endocrinol 2014(1):7. doi:10.1186/1687-9856-2014-7

Bispham J, Budge H, Mostyn A, Dandrea J, Clarke L, Keisler DH, Symonds ME, Stephenson T (2002)

Ambient temperature, maternal dexamethasone, and postnatal ontogeny of leptin in the neonatal lamb. Pediatr Res 52(1):85–90

Blüher S, Panagiotou G, Petroff D, Markert J, Wagner A, Klemm T, Filippaios A, Keller A, Mantzoros CS (2014) Effects of a 1-year exercise and lifestyle intervention on irisin, adipokines, and inflammatory markers in obese children. Obesity (Silver Spring) 22(7):1701–1708

Boström P, Wu J, Jedrychowski MP, Korde A, Ye L, Lo JC, Rasbach KA, Boström EA, Choi JH, Long JZ, Kajimura S, Zingaretti MC, Vind BF, Tu H, Cinti S, Højlund K, Gygi SP, Spiegelman BM (2012) A PGC1-α-dependent myokine that drives brown-fat-like development of white fat and thermogenesis. Nature 481(7382):463–468

Cannon B, Nedergaard J (2004) Brown adipose tissue function and physiological significance. Physiol Rev 84(1):277–359

Clarke L, Bryant MJ, Lomax MA, Symonds ME (1997a) Maternal manipulation of brown adipose tissue and liver development in the ovine fetus during late gestation. Br J Nutr 77(6):871–883

Clarke L, Buss DS, Juniper DT, Lomax MA, Symonds ME (1997b) Adipose tissue development during early postnatal life in ewe-reared lambs. Exp Physiol 82(6):1015–1027

Clarke L, Heasman L, Symonds ME (1998) Influence of maternal dexamethasone administration on thermoregulation in lambs delivered by caesarean section. J Endocrinol 156(2):307–314

Cypess AM, Lehman S, Williams G, Tal I, Rodman D, Goldfine AB, Kuo FC, Palmer EL, Tseng YH, Doria A (2009) Identification and importance of brown adipose tissue in adult humans. N Engl J Med 360:1509–1517

Drubach LA, Palmer EL, Connolly LP, Baker A, Zurakowski D, Cypess AM (2011) Pediatric brown adipose tissue: detection, epidemiology, and differences from adults. J Pediatr 159(6):939–944

Gemmell RT, Alexander G (1978) Ultrastructural development of adipose tissue in foetal sheep. Aust J Biol Sci 31(5):505–515

Gemmell RT, Bell AW, Alexander G (1972) Morphology of adipose cells in lambs at birth and during subsequent transition of brown to white adipose tissue in cold and in warm conditions. Am J Anat 133(2):143–164

Giannini C, Feldstein AE, Santoro N, Kim G, Kursawe R, Pierpont B, Caprio S (2013) Circulating levels of FGF-21 in obese youth: associations with liver fat content and markers of liver damage. J Clin Endocrinol Metab 98(7):2993–3000

Gilsanz VS, Chung A, Jackson H, Dorey FJ, Hu HH (2011) Functional brown adipose tissue is related to muscle volume in children and adolescents. J Pediatr 158(5):722–726

Gnanalingham MG, Mostyn A, Symonds ME, Stephenson T (2005) Ontogeny and nutritional programming of adiposity in sheep: potential role of glucocorticoid

action and uncoupling protein-2. Am J Physiol 289 (5):R1407–R1415

Gregoire FM, Smas CM, Sul HS (1998) Understanding adipocyte differentiation. Physiol Rev 78(2):783–809

Hall JA, Ribich S, Christoffolete MA, Simovic G, Correa-Medina M, Patti ME, Bianco AC (2010) Absence of thyroid hormone activation during development underlies a permanent defect in adaptive thermogenesis. Endocrinology 151(9):4573–4582

Heasman L, Clarke L, Symonds ME (2000) Influence of thyrotropin-releasing hormone administration at birth on thermoregulation in lambs delivered by cesarean. Am J Obstet Gynecol 183(5):1257–1262

Hondares E, Rosell M, Gonzalez FJ, Giralt M, Iglesias R, Villarroya F (2010) Hepatic FGF21 expression is induced at birth via PPARalpha in response to milk intake and contributes to thermogenic activation of neonatal brown fat. Cell Metab 11(3):206–212

Huh JY, Panagiotou G, Mougios V, Brinkoetter M, Vamvini MT, Schneider BE, Mantzoros CS (2012) FNDC5 and irisin in humans: I. Predictors of circulating concentrations in serum and plasma and II. MRNA expression and circulating concentrations in response to weight loss and exercise. Metabolism 61(12):1725–1738

Itoh N (2014) FGF21 as a hepatokine, adipokine, and myokine in metabolism and diseases. Front Endocrinol (Lausanne) 5:107. doi:10.3389/fendo. 2014.00107

Kajimura S, Seale P, Kubota K, Lunsford E, Frangioni JV, Gygi SP, Spiegelman BM (2009) Initiation of myoblast to brown fat switch by a PRDM16-C/EBP-β transcriptional complex. Nature 460 (7259):1154–1158

Lo KA, Sun L (2013) Turning WAT into BAT: a review on regulators controlling the browning of white adipocytes. Biosci Rep 33(5):pii:e00065. doi:10. 1042/BSR20130046

Mostyn A, Bispham J, Pearce S, Evens Y, Raver N, Keisler DH, Webb R, Stephenson T, Symonds ME (2002) Differential effects of leptin on thermoregulation and uncoupling protein abundance in the neonatal lamb. FASEB J 16(11):1438–1440

Mostyn A, Pearce S, Budge H, Elmes M, Forhead AJ, Fowden AL, Stephenson T, Symonds ME (2003) Influence of cortisol on adipose tissue development in the fetal sheep during late gestation. J Endocrinol 176(1):23–30

Mostyn A, Attig L, Larcher T, Dou S, Chavatte-Palmer P, Boukthir M, Gertler A, Djiane J, Symonds ME, Abdennebi-Najar L (2014) UCP1 is present in porcine adipose tissue and is responsive to postnatal leptin. J Endocrinol 223(1):M31–M38

Norheim F, Langleite M, Hjorth M, Holen T, Kielland A, Stadheim HK, Gulseth HL, Birkeland KI, Jensen J, Drevon CA (2014) The effects of acute and chronic exercise on PGC-1α, irisin and browning of subcutaneous adipose tissue in humans. FEBS J 281(3):739–749

Ohta H, Itoh N (2014) Roles of FGFs as adipokines in adipose tissue development, remodeling, and metabolism. Front Endocrinol (Lausanne) 5:18. doi:10.3389/fendo.2014.0001818

Ornoy A (2011) Prenatal origin of obesity and their complications: gestational diabetes, maternal overweight and the paradoxical effects of fetal growth restriction and macrosomia. Reprod Toxicol 32:205–212

Pan D, Fujimoto M, Lopes A, Wang YX (2009) Twist-1 is a PPARdelta-inducible, negative-feedback regulator of PGC-1alpha in brown fat metabolism. Cell 137(1):73–86

Pearce S, Budge H, Mostyn A, Genever E, Webb R, Ingleton P, Walker AM, Symonds ME, Stephenson T (2005) Prolactin, the prolactin receptor and uncoupling protein abundance and function in adipose tissue during development in young sheep. J Endocrinol 184(2):351–359

Pope M, Budge H, Symonds ME (2014) The developmental transition of ovine adipose tissue through early life. Acta Physiol (Oxf) 210(1):20–30

Raschke S, Elsen M, Gassenhuber H, Sommerfeld M, Schwahn U, Brockmann B, Jung R, Wisløff U, Tjønna AE, Raastad T, Hallén J, Norheim F, Drevon CA, Romacho T, Eckardt K, Eckel J (2013) Evidence against a beneficial effect of irisin in humans. PLoS One 8(9):e73680. doi:10.1371/journal.pone.0073680

Ribeiro MO, Bianco SD, Kaneshige M, Schultz JJ, Cheng SY, Bianco AC, Brent GA (2010) Expression of uncoupling protein 1 in mouse brown adipose tissue is thyroid hormone receptor-beta isoform specific and required for adaptive thermogenesis. Endocrinology 151(1):432–440

Saito M, Okamatsu-Ogura Y, Matsushita M, Watanabe K, Yoneshiro T, Nio-Kobayashi J, Iwanaga T, Miyagawa M, Kameya T, Nakada K (2009) High incidence of metabolically active brown adipose tissue in healthy adult humans: effects of cold exposure and adiposity. Diabetes 58:1526–1531

Scholzen T, Gerdes J (2000) The Ki-67 protein: from the known and the unknown. J Cell Physiol 182(3):311–322

Schulz TJ, Tseng YH (2009) Emerging role of bone morphogenetic proteins in adipogenesis and energy metabolism. Cytokine Growth Factor Rev 20 (5–6):523–531

Seale P (2010) Transcriptional control of brown adipocyte development and thermogenesis. Int J Obes 34(Suppl 1):S17–S22

Staiger H, Böhm A, Scheler M, Berti L, Machann J, Schick F, Machicao F, Fritsche A, Stefan N, Weigert C, Krook A, Häring HU, de Angelis MH (2013) Common genetic variation in the human FNDC5 locus, encoding the novel muscle-derived 'browning' factor irisin, determines insulin sensitivity. PLoS One 8(4):e61903. doi:10.1371/journal.pone.0061903

Symonds ME (2013) Brown adipose tissue growth and development. Scientifica (Cairo) 2013:305763. doi:10.1155/2013/305763

Symonds ME, Andrews DC, Johnson P (1989) The control of thermoregulation in the developing lamb during slow wave sleep. J Dev Physiol 11(5):289–298

Symonds ME, Bird JA, Clarke L, Gate JJ, Lomax MA (1995) Nutrition, temperature and homeostasis during perinatal development. Exp Physiol 80(6):907–940

Symonds ME, Bird JA, Sullivan C, Wilson V, Clarke L, Stephenson T (2000) Effect of delivery temperature on endocrine stimulation of thermoregulation in lambs born by cesarean section. J Appl Physiol 88(1):47–53

Symonds ME, Mostyn A, Pearce S, Budge H, Stephenson T (2003) Endocrine and nutritional regulation of fetal adipose tissue development. J Endocrinol 179(3):293–299

Taittonen M, Laine J, Savisto N (2009) Functional brown adipose tissue in healthy adults. N Engl J Med 360:1518–1525

Tews D, Wabitsch M (2011) Renaissance of brown adipose tissue. Horm Res Paediatr 75(4):231–239

Timmons JA, Wennmalm K, Larsson O, Walden TB, Lassmann T, Petrovic N, Hamilton DL, Gimeno RE, Wahlestedt C, Baar K, Nedergaard J, Cannon B (2007) Myogenic gene expression signature establishes that brown and white adipocytes originate from distinct cell lineages. Proc Natl Acad Sci U S A 104:4401–4406

Timmons JA, Baar K, Davidsen PK, Atherton PJ (2012) Is irisin a human exercise gene? Nature 488(7413): E9–E10

Tseng YH, Kokkotou E, Schulz TJ, Huang TL, Winnay JN, Taniguchi CM, Tran TT, Suzuki R, Espinoza DO, Yamamoto Y, Ahrens MJ, Dudley AT, Norris AW, Kulkarni RN, Kahn CR (2008) New role of bone morphogenetic protein 7 in brown adipogenesis and energy expenditure. Nature 454:1000–1004

Uldry M, Yang W, St-Pierre J, Lin J, Seale P, Spiegelman BM (2006) Complementary action of the PGC-1 coactivators in mitochondrial biogenesis and brown fat differentiation. Cell Metab 3(5):333–341

Van Marken Lichtenbelt WD, Vanhommerig JW, Smulders NM, Drossaerts JM, Kemerink GJ, Bouvy ND, Schrauwen P, Teule GJ (2009) Cold-activated brown adipose tissue in healthy men. N Engl J Med 360:1500–1508

Viengchareun S, Servel N, Fève B, Freemark M, Lombès M, Binart N (2008) Prolactin receptor signaling is essential for perinatal brown adipocyte function: a role for insulin-like growth factor-2. PLoS One 3(2): e1535. doi:10.1371/journal.pone.0001535

Virtanen KA, Lidell ME, Orava J, Heglind M, Westergren R, Niemi T, Taittonen M, Laine J, Savisto NJ, Enerback S, Nuutila P (2009) Functional brown adipose tissue in healthy adults. N Engl J Med 360(15):1518–1525

Wrann CD, White JP, Salogiannnis J, Laznik-Bogoslavski D, Wu J, Ma D, Lin JD, Greenberg ME, Spiegelman BM (2013) Exercise induces hippocampal BDNF through a PGC-1α/FNDC5 pathway. Cell Metab 18(5):649–659

Wu J, Boström P, Sparks LM, Ye L, Choi JH, Giang AH, Khandekar M, Virtanen KA, Nuutila P, Schaart G, Huang K, Tu H, van Marken Lichtenbelt WD, Hoeks J, Enerbäck S, Schrauwen P, Spiegelman BM (2012) Beige adipocytes are a distinct type of thermogenic fat cell in mouse and human. Cell 152(2):366–376

Wu J, Cohen P, Spiegelman BM (2013) Adaptive thermogenesis in adipocytes: is beige the new brown? Genes Dev 27:234–250

Yu ZB, Han SP, Zhu GZ, Zhu C, Wang XJ, Cao XG, Guo XR (2011) Birth weight and subsequent risk of obesity: a systematic review and meta-analysis. Obes Rev 12:525–542

Zingaretti MC, Crosta F, Vitali A, Guerrieri M, Frontini A, Cannon B, Nedergaard J, Cinti S (2009) The presence of UCP1 demonstrates that metabolically active adipose tissue in the neck of adult humans truly represents brown adipose tissue. FASEB J 23(9):3113–3120

Advs Exp. Medicine, Biology - Neuroscience and Respiration (2015) 15: 35–40
DOI 10.1007/5584_2015_147
© Springer International Publishing Switzerland 2015
Published online: 29 May 2015

Regulatory T Cells in Obesity

Anna M. Kucharska, Beata Pyrżak, and Urszula Demkow

Abstract

The current concept of the pathogenesis of obesity relates to the inflammation caused by excess of adipose tissue. Regulatory T cells accumulated in visceral adipose tissue (VAT-resident Tregs) are also involved in this pathogenesis. In the present paper the mechanisms responsible for alterations in the number and function of VAT-resident Tregs T in obesity are described. The role of Tregs in inflammation, insulin resistance, atherogenesis, and also the influence on VAT-resident Tregs of adipocytokines and insulin are reviewed.

Keywords

Immune regulation • Insulin • Leptin • Obesity • Tregs

1 Introduction

In the last decades obesity became the leading health problem in developed countries. Its frequency increases significantly in young people and in pediatric population (de Onis et al. 2010). It is well known that obesity concerns not only the metabolism but influences endocrine, and immune and cardiovascular systems. The present concept of obesity-induced comorbidities relates

A.M. Kucharska (✉) and B. Pyrżak
Department of Pediatrics and Endocrinology, Medical University of Warsaw, 24 Marszalkowska St., 00-576 Warsaw, Poland
e-mail: anna.kucharska@litewska.edu.pl

U. Demkow
Department of Laboratory Diagnostics and Clinical Immunology of the Developmental Age,
Medical University of Warsaw, Warsaw, Poland

to the inflammation caused by the excess of adipose tissue (de Jong et al. 2014). It is well documented that the inflammation is strongly associated with insulin resistance and atherosclerosis in obese patient and leads to type 2 diabetes (Nikolajczyk et al. 2012).

Adipose tissue consists mainly of adipocytes demonstrating the ability to store triglycerides and to regulate energy metabolism by secretion of metabolic hormones such as leptin and adiponectin. Additionally, the adipose tissue contains a considerable number of macrophages, neutrophils, T cells, B cells, invariant natural killer cells, eosinophils, and mast cells (Gregor and Hotamisligil 2011; Anderson et al. 2010). The number of macrophages positively correlates with body mass index (Weisberg et al. 2003) and the increase in adipose tissue correlates with the macrophage proinflammatory phenotype M1 (Lumeng et al. 2007). Taking into account that adipose tissue in obese individuals contains a high number of cells different from adipocytes and the majority of them are recognized as important in immune response, it is clear that the excess of adipose tissue would have immunological implications and would influence the function of true adipocytes. Macrophages are reported as the key players in the development of obesity-associated inflammation (Lumeng et al. 2007), but recent observations suggest a considerable contribution of regulatory T cells (Tregs) in this process (Matarese et al. 2010a; Lolmède et al. 2011). In the course of obesity, a progressive infiltration of visceral adipose tissue (VAT) with immune cells becomes apparent. This phenomenon is caused by macrophage activation induced by adipocyte death (Chen et al. 2013). The expansion of effector T cells and increases in the number of B-2 cells, T helper type 1 cells, and CD8+ cytotoxic T cells producing proinflammatory cytokines – interferon-γ, tumor necrosis factor, and IL-6 – has been observed in VAT (Chen et al. 2013). Recent studies confirm the notion that in slim individuals VAT contains Tregs and their number increases with age, but decreases in obesity (Feuerer et al. 2009; Winer et al. 2009; Ilan et al. 2010). This observations highlight a close association between obesity, chronic inflammation, and the Tregs accumulation in adipose tissue.

2   Tregs Residing in Visceral Adipose Tissue

The origin of visceral adipose tissue Treg cells (VAT-resident Tregs) is not well defined. The current model of the Treg development proposes that 'natural' Tregs are generated in the thymus and migrate to peripheral tissues However, it has also been reported that Tregs can be induced from peripheral naive CD4+ T lymphocytes during immune responses, depending on the antigen concentration and the cytokine milieu (Sakaguchi et al. 2008). Natural Tregs belong to the subset of CD4+ cells, positive for CD25 and expressing transcription factor fork-head box P3 (FoxP3),

cytotoxic T-lymphocyte antigen 4 (CTLA-4), and glucocorticoid-induced tumor necrosis factor receptor (GITR). They play an essential role in the maintaining of self-tolerance and the control of immune responses (Sakaguchi et al. 2008). The mechanisms of Treg action are not well understood. Nevertheless it is established that Tregs are inhibitors of immune responses and their deficiency leads to autoimmune or allergic diseases. Recent studies have shown that Tregs may also be involved in the pathogenesis of obesity-associated inflammation and insulin resistance (Cipolletta et al. 2011; Schipper et al. 2012; Winer and Winer 2012. Wagner et al. 2013). Tregs constitute a small number of the peripheral T cells and count for around 5–15 % of the general CD4+ T cell subset. However, it has been recently reported that the subset of CD4 + FoxP3+ Tregs is present in a greater amount in VAT and these cells accumulate with age in lean mice (Feuerer et al. 2009). The relative number of Tregs among CD4+ T cells is markedly greater in VAT than in other tissues in healthy individuals and the number declines in obesity. A significant reduction of Tregs in VAT is strongly correlated with inflammation and insulin resistance (Eller et al. 2011; Feuerer et al. 2009). It has been reported in mice on a high fat diet that the expanded population of Tregs has a protective role against metabolic disorders (Pettersson et al. 2012). In obese mice, a proinflammatory T cell profile including increased Th17 cells associated with a decreased number of Tregs is present (Feuerer et al. 2009; Yang et al. 2010). Similar relations exist in patients with type 2 diabetes (DeFuria et al. 2013). Recently, a new subpopulation of Tregs, known as visceral adipose-tissue-resident Foxp3 + CD4+ T cells (VAT Tregs) has been described, with important implications for the pathogenesis of obesity and its metabolic consequences.

3   VAT Tregs in Obesity-Associated Inflammation and Metabolic Disorders

The influence of VAT Tregs on metabolic homeostasis in obesity is an area of limited understanding. It is known that the presence of VAT

Tregs is associated with a lower degree of inflammation and better insulin sensitivity (Lumeng and Saltiel 2011). In non-obese individuals, the percentage of Tregs in the CD4+ subset is greater in visceral than subcutaneous adipose tissue, or in other compartments such as lymph nodes, spleen, and non-lymphoid tissues. In obese individuals, the number of Tregs decreases significantly only in VAT (Deiuliis et al. 2011). This reduction is strongly associated with an increment of inflammatory mediators in VAT and a decrease of insulin sensitivity (Eller et al. 2011). Conversely, Tregs transfer or induction of Tregs *in vivo* is combined with significant improvement of glucose tolerance and insulin sensitivity in obese mice (Ilan et al. 2010; Eller et al. 2011). Similar effects have been reported by Łuczynski et al. (2012) in children with metabolic syndrome. Insulin resistance is associated with a decrease of VAT Tregs and with adipocytes hypertrophy. Some studies analyzed the possible direct influence of Tregs on adipocyte size. This issue has not yet been resolved, but there are reports of a significant reduction of the adipocyte size after the induction of Tregs in mice (Tian et al. 2011). However, other authors report no impact of induced or transferred Tregs on the adipocytes size or number (Ilan et al. 2010; Eller et al. 2011).

Tregs are also reported as protective against atherosclerosis. Atherosclerotic plaque formation is mediated by Th1 cells and Tregs counteract the autoimmune reaction evoked by Th1 against oxidized LDL. In db/db mice, supplementation of deficient Treg cells reduces the size of atherosclerotic lesions and significantly inhibits IFN-γ production (Taleb et al. 2008). Moreover, it is shown that leptin/leptin receptor signaling improves Treg function and protects from atherosclerosis in experimental models (Taleb et al. 2007).

The molecular mechanisms underlying the T-cell-mediated inflammation in adipose tissue are the matter of intensive studies. Peroxisome proliferator-activated receptor gamma (PPAR-γ) is considered the crucial molecular factor controlling VAT Treg accumulation, phenotypes, and functions (Cipolletta et al. 2011). In slim subjects, PPAR-γ is responsible for maintaining the unique phenotype of VAT Tregs, notably concerning chemokine receptors, which facilitates Treg recruitment into VAT. PPAR-γ activation can promote adipocytes to secret adiponectin, which is crucial for VAT Treg accumulation by inducing IL-10 synthesis and promoting M2 polarization of macrophages (Ouchi et al. 2011). In obese subjects, phosphorylation or other post-translational modifications of PPAR-γ may contribute to the reduction of VAT Tregs by altering their unique phenotype (Deiuliis et al. 2011). Additionally, phosphorylation of PPAR-γ blocks adiponectin production in adipocytes and may promote proinflammatory M1 macrophage generation (Choi et al. 2010). Proinflammatory cytokines, such as IFN-γ, TNF-α, IL-1, and leptin released from adipocytes or M1 macrophages inhibit VAT Treg accumulation.

Another important molecular mechanism of VAT Tregs regulation in obesity is the signal transducer and activator of transcription 3 (STAT3) (Priceman et al. 2013; Durant et al. 2010). STAT3 activity is increased in obese VAT and in VAT-resident T cells. Functional ablation of STAT3 in T cells in diet-induced obesity improves insulin sensitivity and glucose tolerance, and suppresses VAT inflammation. Moreover, ablation of STAT3 reverses the increased Th1/Treg ratio in VAT obtained from diet-induced obese mice (Priceman et al. 2013). This phenomenon might be secondary to the enhanced IL-6 production. Such processes lead to suppression of VAT Tregs. Additionally, STAT3 in T cells in diet-induced obese mice supports VAT macrophage accumulation and induces a shift to M2 phenotype. Therefore, STAT3 in VAT-resident T cells can be considered the influential regulator of adipose tissue inflammation and associated metabolic dysfunctions (Priceman et al. 2013).

4 Leptin Influence on Treg Cells

Leptin plays a role in Tregs formation and modifies their proliferative capacity (De Rosa et al. 2007). Chronic leptin and leptin receptor deficiency is linked with an increased percentage of Tregs, and their higher activity and resistance to autoimmune diseases. This effect can be reversed by leptin substitution (Matarese et al. 2001, 2010b). In humans, leptin decreases the proliferation of Tregs. It is well documented by *in vitro* studies that leptin neutralization following anti-CD3/CD28 antibody stimulation increases Tregs proliferation (De Rosa et al. 2007). In mice, leptin and leptin receptor deficiency results in an increased number of Tregs and their activity (Matarese et al. 2001). It has been confirmed that Tregs express leptin receptor and are able to secrete leptin (Papathanassoglou et al. 2006). Freshly isolated Tregs produce a high amount of leptin and express the long signaling form of the leptin receptor (De Rosa et al. 2007; Conde et al. 2010, 2014).

In patients with multiple sclerosis, the number of peripheral Tregs is inversely correlated with the serum leptin level (Matarese et al. 2010a). In obesity, serum leptin is elevated and can also influence circulating Tregs, but the VAT Tregs cross-talking with adipocytes might be much more important for maintaining the metabolic balance. Tregs accumulate preferentially in lean tissues, thus in obesity their number is reduced (Feuerer et al. 2009). Leptin has opposite effects on Th1 and Tregs. Adipocytes-derived leptin stimulates Th1 cells to proliferate and release proinflammatory cytokines, but inhibits the proliferation of Tregs. Treg cells also secret leptin and in this autocrine loop control their own responsiveness. Leptin neutralization (both adipocytes-derived and Treg-derived) can reverse this effects (De Rosa et al. 2007).

The basic mechanism of a leptin inhibitory effect on Treg proliferation is the leptin-mTOR signaling pathway (Procaccini et al. 2010, 2012). Leptin inhibits rapamycin- induced proliferation of Tregs. An opposite effect is observed in leptin receptor-deficient mice; decreased mTOR activity in Tregs and increased proliferation of Tregs (Procaccini et al. 2010). A recently published study by Moraes-Vieira et al. (2014) has shown that leptin can control Tregs indirectly *via* dendritic cells. These authors observed that leptin deficiency reduces the level of markers of

dendritic cells maturation and decreases production of proinflammatory cytokines (IL-12, TNF-α, and IL-6) in association with increased production of TGF-β by Tregs. This unique phenotype of dendritic cells supports the generation of induced Treg or TH17 cells in leptin-free conditions more efficiently than in the presence of leptin. The results indicate that in the absence of leptin Treg-cell induction requires dendritic cells with increased TGF-β expression. In aggregate, it might be possible that dendritic cells activated in the absence of leptin are responsible for mTOR activation in CD4+ T cells and in this way dendritic cells may account for specific T cells transformations.

5 Adiponectin Influence on Tregs

A direct effect of adiponectin on VAT Tregs has not yet been studied. Nevertheless, adiponectin is highly expressed in visceral opposite to subcutaneous adipose-tissue. Adiponectin can induce IL-10 synthesis and switches the macrophage phenotype to anti-inflammatory M2 (Ouchi et al. 2011; Wolf et al. 2004). Therefore, adiponectin may contribute indirectly to Tregs induction in VAT. IL-10 is secreted by macrophages M2, depending on adiponectin stimulation. The presence of considerable amounts of IL-10 supports the maintaining of FoxP3 expression and activity of Tregs in VAT (Lumeng et al. 2007).

6 Insulin Influence on Tregs

The association between a low number of Tregs and insulin resistance in obesity has been described by many authors (Ilan et al. 2010; Eller et al. 2011; Cipolletta et al. 2011), but the precise mechanism of direct influence of insulin on Tregs remains obscure. Han et al. (2014) have confirmed that insulin controls Tregs function *via* IL-10. In mice, regulatory T cells express the insulin receptor. High concentrations of insulin impair the ability of Tregs to suppress inflammatory responses *via* effects on the AKT/mTOR signaling pathway. Insulin-activated AKT signaling in Tregs suppresses both IL-10 production and the ability of Tregs to suppress the production of TNF-α by macrophages in a contact-independent manner. In mice with diet-induced obesity, Tregs obtained from the visceral adipose tissue produce low amounts of IL-10 accompanied by an increase in the production of IFN-γ. These data suggest that high levels of insulin may promote the development of obesity-associated inflammation and insulin resistance *via* the effect on the IL-10–mediated function of Tregs (Han et al. 2014).

In conclusion, a better understanding of immunological dysfunctions in obesity opens up new perspectives for therapy and prevention of obesity and obesity-induced comorbidities.

Conflicts of Interest The authors declare no conflicts of interest in relation to this article.

References

Anderson EK, Gutierrez DA, Hasty AH (2010) Adipose tissue recruitment of leukocytes. Curr Opin Lipidol 21:172–177

Chen X, Wu Y, Wang L (2013) Fat-resident Tregs: an emerging guard protecting from obesity-associated metabolic disorders. Obes Rev 14:568–578

Choi JH, Banks AS, Estall JL, Kajimura S, Boström P, Laznik D, Ruas JL, Chalmers MJ, Kamenecka TM, Blüher M, Griffin PR, Spiegelman BM (2010) Antidiabetic drugs inhibit obesity-linked phosphorylation of PPAR-gamma by Cdk5. Nature 466 (7305):451–456

Cipolletta D, Kolodin D, Benoist C, Mathis D (2011) Tissular T(regs): a unique population of adipose tissue-resident Foxp3+CD4+ T cells that impacts organismal metabolism. Semin Immunol 23 (6):431–437

Conde J, Scotece M, Gomez R, Gómez-Reino JJ, Lago F, Gualillo O (2010) At the crossroad between immunity and metabolism: focus on leptin. Expert Rev Clin Immunol 6(5):801–808

Conde J, Scotece M, Abella V, Lopez V, Pino J, Gomez-Reino J, Gualillo O (2014) An update on leptin as immunomodulator. Expert Rev Clin Immunol 10 (9):1165–1170

DeFuria J, Belkina AC, Jagannathan-Bogdan M, Snyder-Cappione J, Carr JD, Nersesova JR, Markham D, Strissel KJ, Watkins AA, Zhu M, Allen J, Bouchard J, Toraldo G, Jasuja R, Obin MS, McDonnell ME, Apovian C, Denis GV, Nikolajczyk

BS (2013) B cells promote inflammation in obesity and type 2 diabetes through regulation of T-cell function and an inflammatory cytokine profile. Proc Natl Acad Sci 110(13):5133–5138 MEDIENCES

Deiuliis J, Shah Z, Shah N, Needleman B, Mikami D, Narula V, Perry K, Hazey J, Kampfrath T, Kollengode M, Sun Q, Satoskar AR, Lumeng C, Moffatt-Bruce S, Rajagopalan S (2011) Visceral adipose inflammation in obesity is associated with critical alterations in T regulatory cell numbers. PLoS One 6 (1):e16376. doi:10.1371/journal.pone.0016376

de Jong A, Kloppenburg M, Toes RE, Joan-Facsinay A (2014) Fatty acids, lipid mediators, and T cell function. Front Immunol 5:483. doi:10.3389/fimmu.2014.00483

de Onis M, Blössner M, Borghi E (2010) Global prevalence and trends of overweight and obesity among preschool children. Am J Clin Nutr 92(5):1257–1264

De Rosa V, Procaccini C, Cali G, Pirozzi G, Fontana S, Zappacosta S, La Cava A, Matarese G (2007) A key role of leptin in the control of regulatory T cell proliferation. Immunity 26:241–255

Durant L, Watford WT, Ramos HL, Laurence A, Vahedi G, Wei L, Takahashi H, Sun HW, Kanno Y, Powrie F, O'Shea JJ (2010) Diverse targets of the transcription factor STAT3 contribute to T cell pathogenicity and homeostasis. Immunity 32(5):605–615

Eller K, Kirsch A, Wolf AM, Sopper S, Tagwerker A, Stanzl U, Wolf D, Patsch W, Rosenkranz AR, Eller P (2011) Potential role of regulatory T cells in reversing obesity-linked insulin resistance and diabetic nephropathy. Diabetes 60(11):2954–2962

Feuerer M, Herrero L, Cipolletta D, Naaz A, Wong J, Nayer A, Lee J, Goldfine AB, Benoist C, Shoelson S, Mathis D (2009) Lean, but not obese, fat is enriched for a unique population of regulatory T cells that affect metabolic parameters. Nat Med 15:930–939

Gregor MF, Hotamisligil GS (2011) Inflammatory mechanisms in obesity. Annu Rev Immunol 29:415–445

Han JM, Patterson SJ, Speck M, Ehses JA, Levings MK (2014) Insulin inhibits IL-10-mediated regulatory T cell function: implications for obesity. J Immunol 192:623–629

Ilan Y, Maron R, Tukpah AM, Maioli TU, Murugaiyan G, Yang K, Wu HY, Weiner HL (2010) Induction of regulatory T cells decreases adipose inflammation and alleviates insulin resistance in ob/ob mice. Proc Natl Acad Sci USA 107(21):9765–9770

Lolmède K, Duffaut C, Zakaroff-Girard A, Bouloumie A (2011) Immune cells in adipose tissue: key players in metabolic disorders. Diabetes Metab 37:283–290

Łuczynski W, Wawrusiewicz-Kurylonek N, Iłendo E, Bossowski A, Głowińska-Olszewska B, Krętowski A, Stasiak-Barmuta A (2012) Generation of functional T-regulatory cells in children with metabolic syndrome. Arch Immunol Ther Exp (Warsz) 60 (6):487–495

Lumeng CN, Saltiel AR (2011) Inflammatory links between obesity and metabolic disease. J Clin Invest 121(6):2111–2117

Lumeng CN, Bodzin JL, Saltiel AR (2007) Obesity induces a phenotypic switch in adipose tissue macrophage polarization. J Clin Invest 117(1):175–184

Matarese G, Di Giacomo A, Sanna V, Lord GM, Howard JK, Di Tuoro A, Bloom SR, Lechler RI, Zappacosta S, Fontana S (2001) Requirement for leptin in the induction and progression of autoimmune encephalomyelitis. J Immunol 166(10):5909–5916

Matarese G, Procaccini C, De Rosa V, Horvath TL, La Cava A (2010a) Regulatory T cells in obesity: the leptin connection. Trends Mol Med 16:247–256

Matarese G, Carrieri PB, Montella S, De Rosa V, La Cava A (2010b) Leptin as a metabolic link to multiple sclerosis. Nat Rev Neurol 6(8):455–461

Moraes-Vieira PM, Larocca RA, Bassi EJ, Peron JP, Andrade-Oliveira V, Wasinski F, Araujo R, Thornley T, Quintana FJ, Basso AS, Strom TB, Camara NO (2014) Leptin deficiency impairs maturation of dendritic cells and enhances induction of regulatory T and Th17 cells. Eur J Immunol 44(3):794–806

Nikolajczyk BS, Jagannathan-Bogdan M, Denis GV (2012) The outliers become a stampede as immunometabolism reaches a tipping point. Immunol Rev 249(1):253–275

Ouchi N, Parker JL, Lugus JJ, Walsh K (2011) Adipokines in inflammation and metabolic disease. Nat Rev Immunol 11(2):85–97

Papathanassoglou E, El-Haschimi K, Li XC, Matarese G, Strom T, Mantzoros C (2006) Leptin receptor expression and signaling in lymphocytes: kinetics during lymphocyte activation, role in lymphocyte survival, and response to high fat diet in mice. J Immunol 176(12):7745–7752

Pettersson US, Waldén TB, Carlsson PO, Jansson L, Phillipson M (2012) Female mice are protected against high-fat diet induced metabolic syndrome and increase the regulatory T cell population in adipose tissue. PLoS One 7(9):e46. doi:10.1371/journal.pone.0046057

Priceman SJ, Kujawski M, Shen S, Cherryholmes GA, Lee H, Zhang C, Kruper L, Mortimer J, Jove R, Riggs AD, Yu H (2013) Regulation of adipose tissue T cell subsets by Stat3 is crucial for diet-induced obesity and insulin resistance. Proc Natl Acad Sci 110(32):13079–13084

Procaccini C, De Rosa V, Galgani M, Abanni L, Calì G, Porcellini A, Carbone F, Fontana S, Horvath TL, La Cava A, Matarese G (2010) An oscillatory switch in mTOR kinase activity sets regulatory T cell responsiveness. Immunity 33(6):929–941

Procaccini C, De Rosa V, Galgani M, Carbone F, Cassano S, Greco D, Qian K, Auvinen P, Calì G, Stallone G, Formisano L, La Cava A, Matarese G (2012) Leptin-induced mTOR activation defines a specific molecular and transcriptional signature controlling CD4+ effector T cell responses. J Immunol 189:2941–2953

Sakaguchi S, Yamaguci T, Nomura T, Ono M (2008) Regulatory T cells and immune tolerance. Cell 133:775–787

Schipper HS, Prakken B, Kalkhoven E, Boes M (2012) Adipose tissue resident immune cells: key players in immunometabolism. Trends Endocrinol Metab 23:407–415

Taleb S, Herbin O, Ait-Oufella H, Verreth W, Gourdy P, Barateau V, Merval R, Esposito B, Clément K, Holvoet P, Tedgui A, Mallat Z (2007) Defective leptin/leptin receptor signaling improves regulatory T cell immune response and protects mice from atherosclerosis. Arterioscler Thromb Vasc Biol 27:2691–2698

Taleb S, Tedgui A, Mallat Z (2008) Regulatory T-cell immunity and its relevance to atherosclerosis. J Intern Med 263(5):489–499

Tian J, Dang HN, Yong J, Chui WS, Dizon MP, Yaw CK, Kaufman DL (2011) Oral treatment with γ-aminobutyric acid improves glucose tolerance and insulin sensitivity by inhibiting inflammation in high fat diet-fed mice. PLoS One 6(9):e25338. doi:10.1371/journal.pone.0025338

Wagner NM, Brandhors G, Czepluch F, Lankeit M, Eberle C, Herzberg S, Faustin V, Riggert J, Oellerich M, Hasenfuss G, Konstantinides S, Schafer K (2013) Circulating regulatory T cells are reduced in obesity and may identify subjects at increased metabolic and cardiovascular risk. Obesity 21:461–468

Weisberg SP, McCann D, Desai M, Rosenbaum M, Leibel RL, Ferrante AW Jr (2003) Obesity is associated with macrophage accumulation in adipose tissue. J Clin Invest 112(12):1796–1808

Winer S, Winer DA (2012) The adaptive immune system as a fundamental regulator of adipose tissue inflammation and insulin resistance. Immunol Cell Biol 90(8):755–762

Winer SY, Chan G, Paltser D, Truong H, Tsui J, Bahrami R, Dorfman Y, Wang J, Zielenski F, Mastronardi F, Maezawa Y, Drucker DJ, Engleman E, Winer D, Dosch HM (2009) Normalization of obesity associated insulin resistance through immunotherapy. Nat Med 15:921–929

Wolf AM, Wolf D, Rumpold H, Enrich B, Tilg H (2004) Adiponectin induces the anti-inflammatory cytokines IL-10 and IL-1RA in human leukocytes. Biochem Biophys Res Commun 323(2):630–635

Yang H, Youm YH, Vandanmagsar B, Ravussin A, Gimble JM, Greenway F, Stephens JM, Mynatt RL, Dixit VD (2010) Obesity increases the production of proinflammatory mediators from adipose tissue T cells and compromises TCR repertoire diversity: Implications for systemic inflammation and insulin resistance. J Immunol 185(3):1836–1845

Advs Exp. Medicine, Biology - Neuroscience and Respiration (2015) 15: 41–49
DOI 10.1007/5584_2015_139
© Springer International Publishing Switzerland 2015
Published online: 28 May 2015

Peroxynitrite in Sarcoidosis: Relation to Mycobacterium Stationary Phase

A. Dubaniewicz, L. Kalinowski, M. Dudziak, A. Kalinowska, and M. Singh

Abstract

There is evidence that the same mycobacterial heat shock proteins (Mtb-HSPs), especially HSP16, the main marker of mycobacteria dormant stage, occur in sarcoid tissues and in circulated immune complexes and prompt the immune responses against the different genetic background, leading to the development of acute sarcoidosis (SA)/Löfgren syndrome, chronic SA, latent tuberculosis (TB), or active TB. In SA there is increased monocytes phagocytic activity, decreased clearance of antigens/immune complexes by monocytes, which are resistant to apoptosis, and decreased serum microbicidal/degradable nitrate/nitrite (NOx) concentration. Reduction in NOx may result from the reaction of NOx with superoxide with subsequent production of peroxynitrite ($ONOO^-$). In this study, therefore, we evaluated NOx and $ONOO^-$ levels in supernatants of peripheral blood mononuclear cells cultures treated with Mtb-HSPs from 20 SA patients, 19 TB patients, and 21 healthy volunteers using Griess and rhodamine fluorescence methods. We found significantly greater NOx and $ONOO^-$ concentrations with/without Mtb-HSPs stimulation in SA and TB patients than in controls. However, there were significantly lower NOx and higher $ONOO^-$ levels after Mtb-HSPs induction in SA than TB patients. In summary, in contrast to active TB, increased $ONOO^-$ concentration may explain the low level of NOx with induction of *M. tuberculosis* genetic dormancy program *via* higher Mtb-HSP16 expression in SA.

A. Dubaniewicz (✉)
Department of Pulmonology, Medical University of Gdansk, 7 Debinki St., 80-211 Gdansk, Poland
e-mail: aduban@gumed.edu.pl

L. Kalinowski
Department of Medical Laboratory Diagnostics, Medical University of Gdansk, 7 Debinki St., 80-211 Gdansk, Poland

M. Dudziak
Department of Noninvasive Cardiac Diagnostic, Medical University of Gdansk, 7 Debinki St., 80-211 Gdansk, Poland

A. Kalinowska
Department of Mechanical and Biomedical Engineering, Massachusetts Institute of Technology, 77 Massachusetts Ave, Cambridge, MA, USA

M. Singh
GBF – German National Center for Biotechnology, and CEO LIONEX GmbH, Mascheroder Weg 1, 38124 Braunschweig, Germany

Keywords

Heat shock protein • Mycobacterium • Nitrate/nitrite • PBMC culture •
Peroxynitrite • Sarcoidosis • Tuberculosis

1   Introduction

Sarcoidosis (SA) is a multisystem disorder of
unknown cause (Chen and Moller 2014;
Maertzdorf et al. 2012; Dubaniewicz 2010;
Oswald-Richter et al. 2010; Hunninghake et al.
1999). In the light of the modified Matzinger's
model of immune response (Matzinger 1994),
human heat shock proteins (HSPs) may be considered as the main 'danger signals' (tissue dam-
age-associated molecular patterns – DAMPs) and
microbial HSPs as the pathogen-associated
molecular patterns (PAMPs) recognized by pattern recognition receptors (PRR) on altered antigen presenting cells. HSPs may induce sarcoid
inflammation in response to both non-infectious
and infectious factors in a genetically
predisposed host.

Evolutionary conserved HSPs are expressed
in all organisms and involved in cytoprotection
during cell stress, such as infections, phagocytosis, cytokines, heat/cold shock, acidity, hypoxia,
and oxidative stress (Niforou et al. 2014).
Regarding non-infectious causes of sarcoidosis,
human HSPs may be released at high levels during chronic low-grade exposure to misfolding
amyloid precursor protein in stressed cells,
phagocyted metal fumes, pesticides, pigments
with/without aluminum in tattoos, as well as
due to heat shock in firefighters. Regarding infectious causes of sarcoid models, low-virulence
strains of *mycobacteria* and *propionibacteria*,
recognized   through   pattern   recognition
receptors, such as Toll-like receptors 2, 4, and
9, receptors for Fc fragments of immunoglobulin
G or receptors for fragments of complement generate increased release of both human and microbial HSPs. High chronic spread of PAMPs/
DAMPs, microbial and human HSPs, presented
in the context of HLA with bystander activation,
alters expression of cytokines, co-stimulatory or

adhesion molecules, and Tregs, apoptosis, oxidative stress, inducing an immune response, all
considered in sarcoidosis (Typiak et al. 2014;
Dubaniewicz 2013). Additionally, due to molecular and functional homology between microbial
and human HSPs as well as among microbial
HSPs, they may cross-react inducing the autoimmunity (Dubaniewicz 2010).

We have shown that the same *M. tuberculosis* Mtb-HSP, occurring in sera and lymph
nodes, induce distinct immune response
depending on a different genetic background
of the host (HLA and non-HLA: *NRAMP1I*)
developing an autoimmune acute sarcoidosis
or Löfgren syndrome, chronic sarcoidosis,
latent stationary phase of tuberculosis infection
(LTBI) or active tuberculosis (TB). Moreover,
we have revealed that Mtb-HSP16 was significantly more frequent than Mtb-HSP70 and Mtb-
HSP65 in precipitated immune complexes in SA
compared with the corresponding levels in
active TB (Dubaniewicz et al. 2006a, b, 2013;
Dubaniewicz 2010;).

Mtb-HSP16, induced by a low level of nitric
oxide (NO) is important in the survival of
phagocyted mycobacteria and accumulate in
the LTBI in one-third of the world's population
(Voskuil et al. 2003; Garbe et al. 1999). On the
other hand, Mtb-hsp16 inhibits apoptosis of
phagocytes, decreasing expression of the
inducible isoform of nitric oxide synthase
(iNOS), and consequently NO production, and
induces 'dormant stage' mycobacteria, and also
other bacteria (e.g., *propionibacteria, corynebacterium,* or *streptomyces*), considered as the
etiological factors of sarcoidosis (Dubaniewicz
et al. 2013). Low nitrate/nitrite (NO_3^-/NO_2^-)
may also result from a rapid reaction of nitric
oxide stable metabolites (NOx) and superoxide
anion, with the production of cytotoxic and
immunogenous   peroxynitrite   ($ONOO^-$),

Table 1 Clinical characteristics of tested groups

	Control subjects n = 21(%)	Tuberculosis n = 19 (%)	Sarcoidosis n = 20 (%)
Age –mean, range (year)	38; 22–54	45; 20–79	40; 23–69
Gender			
Female	5 (24)	8 (42)	6 (30)
Male	16 (76)	11 (58)	14 (70)
BCG vaccination	21(100)	19 (100)	20 (100)
Positive PPD skin test	0	10 (100)	0
Recidivans	0	0	0
Symptoms			
Cough	0	19 (100)	11 (55)
Dyspnea	0	19 (100)	8 (40)
Fever	0	19 (100)	4 (20)
Night sweats	0	19 (100)	0
Weight loss	0	19 (100)	0
Erythema nodosum	0	0	10 (50)
Löfgren syndrom	0	0	0

BCG Bacille de Calmette et Guérin, *PPD* purified protein derivative tuberculin skin test

leading to the development of chronic inflammation and autoimmunity being at the core of a host of pathologies diseases (Liaudet et al. 2009; Szabo et al. 2007; Klotz et al. 2002; Bingisser et al. 1998; Grune et al. 1998). To the best our knowledge, there are no studies on peroxynitrite in the pathogenesis of sarcoidosis. Therefore, in the present study we evaluated NOx and ONOO⁻ concentrations in cultures of peripheral blood mononuclear cells (PBMC) treated with Mtb-HSP, which were obtained from sarcoidosis and TB patients.

2 Methods

2.1 Study Populations

Ethical approval for the study was obtained from the Independent Bioethics Committee for Scientific Research, Medical University of Gdansk, Poland. The informed consent was obtained from both patients and controls. Comparative characteristics of patients with sarcoidosis, patients with tuberculosis, and controls are shown in Table 1.

Patients with Sarcoidosis Twenty patients with newly detected pulmonary sarcoidosis were included in the study. Sarcoidosis was diagnosed on the basis of histological (scalenobiopsy of lymph nodes), clinical, and radiological examinations. High resolution computed tomography was used to diagnose Stage I (bilateral hilar lymphadenopathy – 10 patients) and Stage II (bilateral hilar lymphadenopathy and parenchymal infiltrations – 10 patients) of sarcoidosis. None of the patients had Löfgren's syndrome or extrapulmonary sarcoidosis. All patients had a negative purified protein derivative skin test. Microbiological examination of lymph nodes and sputum samples revealed no acid-fast bacilli or fungi (data not shown). The PCR and culture were used to exclude *M. tuberculosis* in tested sarcoid tissues.

Patients with Tuberculosis Nineteen patients with newly detected pulmonary tuberculosis were included in the study. The diagnosis of TB was established using standard clinical, radiographic, and bacteriological criteria (demonstration of acid-fast bacilli in sputum smears and positive sputum culture of *M. tuberculosis*). A positive purified protein derivative skin test was an additional criterion for including in the study group. None of the patients showed the involvement of hilar lymph nodes or extrapulmonary tuberculosis.

Control Group The control group consisted of 21 unrelated healthy blood donor volunteers (aged 22–54 years; 5 women, 16 men). All showed normal results of chest radiographs, blood analysis, and no acid-fast bacilli in sputum smears or sputum culture of the *M. tuberculosis* strain. None of the control subjects showed a positive result for the purified protein derivative skin-test. None of the control or sarcoidosis subjects had a familial history of tuberculosis, sarcoidosis, or autoimmune diseases. All study participants were not infected with HIV.

2.2   Methodology

The investigation was conducted before the initiation of treatment in both patient groups. Venous blood samples (15 ml) were collected into tubes with anticoagulant (Becton Dickinson Company, USA) and supernatant of peripheral blood mononuclear cells (PBMC) was aliquoted and stored at −70 °C until use.

Isolation of PBMC and Cell Culture Peripheral blood mononuclear cells were isolated from the peripheral blood samples, by Ficoll-Paque gradient centrifugation. PBMC, lymphocytes, and monocytes were all taken together to mimic conditions occurring *in vivo*, including possible cell-to-cell interactions between different PBMC subpopulations. After two washings with phosphate buffered saline, PBMC were suspended in RPMI 1640 culture medium supplemented with 5 % fetal calf serum (FCS); 1×10^6 cells were then diluted with 1 ml medium (RPMI 1640 + 5 % fetal bovine serum (FBS)) and cultured in triplicate in plastic 24-well plates (Giboc, Invitrogen Cell Culture; Carlsbad, CA). The cultures were stimulated with 1 μg/ml of phytohemagglutinin or 1 μg/ml of Mtb-HSP (recombinant *M. tuberculosis* HSP70 – batch: L70-01-2, lot 1, *M. tuberculosis* HSP65 – batch: L65-99-1, lot 6, and *M. tuberculosis* HSP16 – batch: L16-00-1, lot 1), antigens were obtained from LIONEX Diagnostics & Therapeutics GmbH, Braunschweig, Germany. All the cultures were incubated for 24 h at 37 °C, in a humidified atmosphere containing 5 % CO_2. The controls consisted of non-stimulated cultures in order to estimate the amplitude of response to applied stimuli and to correct for the non-specific binding of the reagent.

Nitrate/Nitrite (NOx) Concentration The measurement of NOx, as an indirect estimation of NO production, was performed in supernatant samples using modified Griess method (Verdon et al. 1995) with some modifications (Dubaniewicz et al. 2013; Kalinowski et al. 2002). Briefly, triplicate samples were incubated for 3 h at 20 °C with glucose-6-phosphate (500 μmol/l), glucose-6-phosphate dehydrogenase (160 U/l), NADH (1 μmol/l), and nitrate reductase (20 U/l) in phosphate buffer (80 mmol/l, pH 7.5). The Griess reaction was then initiated by the addition of sulfanilamide to a final concentration of 0.5 % wt/vol, orthophosphoric acid (1.25 % vol/vol), and N-(1-naphthyl)ethylenediamine hydrochloride (0.05 %, wt/vol). After further incubation at 20 °C for 10 min, the absorbance of each sample mixture was measured at 540 nm and corrected for opacity by measuring the absorbance at 750 nm. The corrected absorbance was interpolated using a standard curve of absorbance plotted against concentration in order to find the concentration of NO_2^- in the sample. As all NO_3^- has already been reduced to NO_2^- by the use of nitrate reductase, this represented the combined concentration of NO_3/NO_2^- in the samples. Results of the NOx assay were expressed as NO_2^- concentration in μmol/l.

Measurement of Peroxynitrite (ONOO⁻) The production of ONOO⁻ was determined by the oxidation of dihydrorhodamine 123 (DHR) to the fluorescent product rhodamine (Briviba et al. 1996). The confluent cells (5×10^5 cells/35-mm dish) were pre-treated with BSO for 18 h, after which they were loaded with fresh medium containing 5 mM DHR and incubated for 2 h at 37 °C to allow the dye to enter the cells and then washed two times with PBS to remove extracellular DHR. Then, cells were trypsinized, washed, and resuspended in 10 mmol/l Tris-HCl/ 25 mmol/l sucrose, pH 7.4, sonicated for 20 s and centrifuged at 10,000 *g* for 15 min at 4 °C. Supernatant fluorescence was measured using a fluorescence spectrophotometer with the excitation and emission wavelengths of 500 and 536 nm, respectively, at room temperature. Authentic rhodamine 123 standards, 0–400 nmol/l, were used to generate a standard curve. All reagents were obtained from Sigma-Aldrich, Poznan, Poland.

2.3 Data Analysis

The Shapiro-Wilk test was used to evaluate normality of distribution of the obtained values. Then, an unpaired t-test followed by Bonferroni *post-hoc* test was used to compare selected pairs of data. Statistical significance was defined as $p \leq 0.05$. Only significant differences are presented in the full form. Statistical analysis was performed with Statistica vs 10.0 (StatSoft, USA) software.

3 Results

3.1 Nitrate/Nitrite Concentration

All NOx concentrations are presented in μmol/l. NOx in the supernatants of unstimulated (control) PBMC cultures did not differ appreciably between SA and TB patients, but it was significantly higher in both patient groups than that in

healthy controls (8.8 ± 1.2 for SA and 9.8 ± 1.2 for TB *vs.* 4.1 ± 1.0 for controls, $p < 0.001$ for both). On the other side, in the stimulated PBMC cultures, NOx was significantly lower in SA than that in TB (32.6 ± 5.5 *vs.* 51.3 ± 6.1 in Mtb-HSP-induced, and 27.6 ± 5.2 *vs.* 44.9 ± 4.2 in PHA-induced cells, respectively), but it remained higher in both patient groups compared with the control (normal) 21.7 ± 4.4 after Mtb-HSP and the 20.7 ± 3.9 after PHA ($p < 0.001$ for all comparisons). There was also a significant difference in NOx concentration after induction of the cells with Mtb-HSP compared with PHA in SA (32.6 ± 5.5 for Mtb-HSP *vs.* 27.6 ± 5.2 for PHA, $p < 0.05$), but not in TB or control subjects (Fig. 1).

3.2 Peroxynitrite Concentration

All ONOO⁻ concentrations are presented in μmol/l. Like NOx, ONOO⁻ in the supernatants of unstimulated (control) PBMC cultures did not differ appreciably between SA and TB patients, but it was significantly higher in both patient groups than that in healthy controls (68.1 ± 10.6 for SA and 65.4 ± 10.9 for TB *vs.* 21.4 ± 5.7). In contrast to NOx, in the stimulated PBMC cultures ONOO⁻ was significantly higher in SA than in TB (284.0 ± 23.2 *vs.* 202.4 ± 8.3 in Mtb-HSP-induced, and

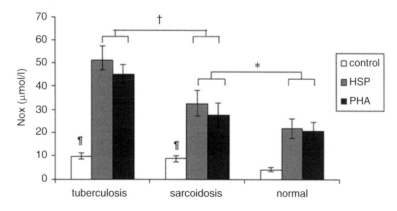

Fig. 1 NOx concentration, as a surrogate measure of NO levels, in PBMCs isolated from healthy controls (normal) and patients with tuberculosis and sarcoidosis. PBMCs were unstimulated or treated with Mtb-HSP (HSP) and PHA. *$p < 0.001$ *vs.* healthy controls with Mtb-HSP and PHA, †$p < 0.001$ *vs.* sarcoidosis with Mtb-HSP and PHA, ¶$p < 0.05$ *vs.* control without Mtb-HSP and PHA (control)

301.5 ± 19.7 *vs.* 210.6 ± 12.4 in PHA-induced cells), but it remained higher in both patient groups compared with the control (normal) 111.0 ± 10.8 after Mtb-HSP and the 113.5 ± 14.5 after PHA (p < 0.001 for all comparisons). In SA, ONOO⁻ concentration was also significantly lower after induction of the cells with Mtb-HSP compared with PHA (284.0 ± 23.2 *vs.* 301.5 ± 19.7, respectively; p < 0.05). In contrast, there were no significant differences in the ONOO⁻ concentration after inductions of the cells with Mtb-HSP or PHA in either healthy subjects or TB patients, although there was a trend toward greater ONOO⁻ concentrations in both groups when the cells were induced with PHA compared with Mtb-HSP (Fig. 2).

4   Discussion

In the present study we found greater increases in NOx and ONOO⁻ content of PBMCs induced by either Mtb-HSPs or HPA in both SA and TB patients compared with those in healthy subjects. However, increases in Mtb-HSP- and HPA-induced NOx production were significantly lower in SA than those in TB patients, whereas increases in Mtb-HSP- and HPA-induced ONOO⁻ production were, conversely,

remarkably higher in SA than those in TB patients. So far, there has been no data on the NOx and ONOO⁻ content in the peripheral blood of patients with SA. There are only a few inconsistent studies on exhaled NO or bronchoalveolar lavage fluid NO content in SA patients (see Dubaniewicz et al. 2013 for a review).

Studies on the genomic and proteomic expression of *mycobacteria* under stress condition, especially a low level of NOx, have shown increased expression of the RNA polymerase sigma (Sig) F unit (formerly *katF*), which is related to the accumulation of Mtb-HSP16 protein, the α-crystallin-related protein, in the cell wall during the bacteria dormant stage (Jee et al. 2008; Garbe et al. 1999). It has been reported that the *katF* gene, more important than the *katG* in turning on transcription of other genes, produces substances which are required for the long-term cell viability (Mulvey et al. 1990). A low level of NOx, as an inhibitor of aerobic respiration in mitochondria inducing the *M. tuberculosis* genetic dormancy program, leads to a non-replicating persistent state characterized by bacteriostasis and structural, metabolic, and chromosal changes in bacteria. Mtb-HSP16 might protect partially unfolded proteins during long periods of bacteriostasis of mycobacteria (Voskuil et al. 2003).

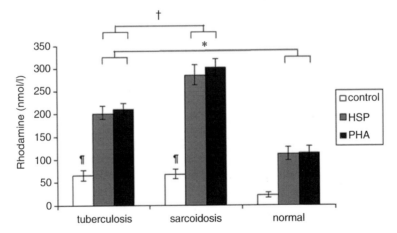

Fig. 2 ONOO⁻ generation by PBMCs isolated from healthy controls (normal) and patients with tuberculosis and sarcoidosis. ONOO⁻ was detected by rhodamine fluorescence in cells untreated or treated Mtb-HSP (HSP) and PHA: *p < 0.001 *vs.* healthy controls with Mtb-HSP and PHA, †p < 0.001 *vs.* sarcoidosis with Mtb-HSP and PHA, ¶p < 0.05 *vs.* healthy controls without Mtb-HSP and PHA

The persistent Mtb-HSP16 stimulation decreases the expression of iNOS, entailing low NO concentration, microbicidal/degradable activity, and apoptosis of mononuclear phagocytes with intracellular mycobacteria chronically releasing HSPs, antigenemia and immunocomplexemia with increased CD4$^+$T- and B-cells proliferation, leading to sarcoid granuloma formation (Arrigo 1998). It is worth noting that especially Mtb-HSP16, having autophosphorylation activity may be involved in increased phagocytosis of *M. bovis* BCG or *M. tuberculosis* by monocytes and macrophages. It is possible that blocking Mtb-HSP16 or protein phosphorylase could be a drug target against infections caused by bacilli and then against sarcoidosis (Prabkhakanar et al. 2000).

On the other hand, a low concentration of NOx may also result from a rapid reaction between NOx and superoxide anion ($O_2{}^{\cdot-}$), with following production of ONOO$^-$, more cytotoxic than NO, which damages cells *via* necrosis or apoptosis, degrades proteins, and increases the load of PAMPs/DAMPs, Ag/immune complexes, including HSPs and cryptogenic Ag, with persistent stimulation of immune cells and induction of immunity (Liaudet et al. 2009; Szabo et al. 2007).

Cellular sources of NO are restricted to various isoforms of NO synthases, whereas $O_2{}^{\cdot-}$ can arise from the mitochondrial electron transport chain, nicotinamide adenine dinucleotide phosphate (NADPH) oxidase, and xanthine oxidase. The rate of reaction $O_2{}^{\cdot-} + \cdot NO \rightarrow ONOO^-$ is at least 3–8 times greater than the rate of $O_2{}^{\cdot-}$ decomposition by superoxide dismutase (SOD), indicating that NO has the ability to drive $O_2{}^{\cdot-}$ away from its main detoxification pathway. Additionally, ONOO$^-$ may also be induced by external factors, e.g., toxins or metals (Liaudet et al. 2009; Szabo et al. 2007). The Mn-cofactored superoxide dismutase peptide A (SodA), toxins, and metals have been considered in the pathogenesis of SA (Dubaniewicz 2013; Oswald-Richter et al. 2010).

Peroxynitrite reacts with a variety of proteins, lipids, and DNA initiating lipid peroxidation, direct inhibition of mitochondrial respiratory chain enzymes, inactivation of glyceraldehyde-3-phosphate dehydrogenase, inhibition of membrane Na$^+$/K$^+$ATP-ase activity, and inactivation of membrane sodium channels. It also damages surfactant proteins, reduces activation of antioxidant enzymes, and activates the precursors of matrix metalloproteinase, which all leads to chronic inflammation, a feature of sarcoidosis (Magri et al. 2013; Piotrowski et al. 2011; Oswald-Richter et al. 2010; Boots et al. 2009; Peltoniemi et al. 2004; Szabo 2003). Moreover, oxidative injury and ATP depletion lead to the activation of the heat shock factor-1, a transcriptional factor which modulates the adaptation and cytoprotective response of the cell through gene expression of HSPs (Andreone et al. 2003). On the other hand, ONOO$^-$ may react with heat shock proteins, e.g., HSP16 or α-crystallins, inducing autoimmune uveitis, cataracta, or glaucoma (Thiagarajan et al. 2004). A high content of ONOO$^-$ is a potent trigger of DNA strand breakage, with the following activation of the nuclear enzyme poly(ADP-ribose) synthetase or polymerase (PARP), with eventual cell energy depletion and necrosis (Szabo 2003). Autoantibodies to poly(ADP-ribose)polymerase were detected in some autoimmune disorders, including SA (van Maarsseveen et al. 2009; Negri et al. 1990). Conversely, reduced content of peroxynitrite may induce apoptosis *via* mitochondrial and DNA injury, 3-, 2-, 8-, and 9-caspase activation, signal transduction disturbances, or dysfunctional calcium pumps. The peroxynitrite-driven apoptotic death and impairment of tyrosine phosphorylation inhibit T lymphocyte activation and proliferation (Dubaniewicz 2013; Brito et al. 1999). ONOO$^-$ also activates cellular immune signaling pathways, among others, *via* NF-κB, mitogen-activated protein kinases, phosphatidylinositol 3-kinases/protein kinase B, or Src family kinases (Liaudet et al. 2009; Klotz et al. 2002; Bingisser et al. 1998). NOx and peroxynitrite change the expression of adhesion molecules, such as ICAM-1, P-selectin, receptor CD11b/CD18, and interleukins and tumor necrosis factor alpha in PBMC; all being altered in both SA and TB (Liaudet et al. 2009; Szabo 2003). Additionally, ONOO$^-$ increases the degradation of cellular proteins by inducing immunoproteasome formation which influences epitope liberation, e. g., of Mtb-HSP16; the processes being considered at play in the pathogenesis of SA (Dubaniewicz 2010; Grune et al. 1998). Moreover, a high degree of homology between bacterial and host HSPs as PAMPs and/or DAMPs, may also play a role in maintaining the latent state of intracellular bacteria, considered in the pathogenesis of sarcoidosis, such as *propionibacterium, corynebacterium,* or *streptomyces* (Jee et al. 2008; Voskuil et al. 2003; Farrar et al. 2000).

In summary, in contrast to active TB, increased immunogenous $ONOO^-$ concentration may explain the low level of NOx with the induction of *M. tuberculosis* genetic dormancy program *via* higher Mtb-HSP16 expression in SA.

Acknowledgements The study was funded by the Ministry of Science and Higher Education and the National Science Center in Poland (grants 5160/B/P01/2010/39 and W-119 to A.D., 2789/B/P01/2010/39 and 6336/B/P01/2011/40 to L.K., and ST-98 to M.D.).

Conflicts of Interest The authors declare no conflicts of interesting in relation to this article.

References

Andreone TL, O'Connor M, Denenberg A, Hake PW, Zingarelli B (2003) Poly(ADP-Ribose) polymerase-1 regulates activation of activator protein-1 in murine fibroblast. J Immunol 170:2113–2120

Arrigo AP (1998) Small stress proteins: chaperones that act as regulators of intracellular redox state and programmed cell death. Biol Chem 379:19–26

Bingisser RM, Tilbrook PA, Holt PG, Kees UR (1998) Macrophage-derived nitric oxide regulates T cell activation via reversible disruption of the Jak3/STAT5 signaling pathway. J Immunol 160:5729–5734

Boots AW, Drent M, Swennen EL, Moonen HJ, Bast A, Haenen GR (2009) Antioxidant status associated with inflammation in sarcoidosis: a potential role for antioxidants. Respir Med 103:364–372

Brito C, Naviliat M, Tiscornia AC, Vuillier F, Gualco G, Dighiero G, Radi R, Cayota AM (1999) Peroxynitrite inhibits lymphocyte activation and proliferation by promoting impairment of tyrosine and phosphorylation and peroxynitrite-driven apoptotic death. J Immunol 162:3356–3366

Briviba K, Roussyn I, Sharov VS, Sies H (1996) Attenuation of oxidation and nitration reactions of peroxynitrite by selenomethionine, selenocystine and ebselen. Biochem J 319:13–15

Chen ES, Moller DR (2014) Etiologic role of infectious agents. Semin Respir Crit Care Med 35:285–289

Dubaniewicz A (2010) Mycobacterium tuberculosis heat shock proteins and autoimmunity in sarcoidosis. Autoimmun Rev 9:419–424

Dubaniewicz A (2013) Microbial and human heat shock proteins as 'danger signals' in sarcoidosis. Hum Immunol 74:1550–1558

Dubaniewicz A, Dubaniewicz-Wybieralska M, Sternau A, Zwolska Z, Izycka-Swieszewska E, Augustynowicz-Kopec E, Skokowski J, Singh M, Zimnoch L (2006a) Mycobacterium tuberculosis complex and mycobacterial heat shock proteins in lymph node tissue from patients with pulmonary sarcoidosis. J Clin Microbiol 44:3448–3451

Dubaniewicz A, Kämpfer S, Singh M (2006b) Serum antimycobacterial heat shock proteins antibodies in sarcoidosis and tuberculosis. Tuberculosis (Edinb) 86:60–67

Dubaniewicz A, Holownia A, Kalinowski L, Wybieralska M, Dobrucki IT, Singh M (2013) Is mycobacterial heat shock protein 16 kDa, a marker of the dormant stage of Mycobacterium tuberculosis, a sarcoid antigen? Hum Immunol 74:45–51

Farrar MD, Ingham E, Holland KT (2000) Heat shock proteins and inflammatory acne vulgaris: molecular cloning, overexpression and purification of a Propionibacterium acnes, GroEL and DnaK homologue. FEMS Microbiol Lett 91:183–186

Garbe TR, Hibler NS, Deretic V (1999) Response to reactive nitrogen intermediates in Mycobacterium tuberculosis: induction of the 16-kilodalton α-crystallin homolog by exposure to nitric oxide donors. Infect Immun 67:460–465

Grune T, Blasig IE, Sitte N, Roloff B, Haseloff R, Davies KJ (1998) Peroxynitrite increases the degradation of aconitase and other cellular proteins by proteasome. J Biol Chem 273:10857–10862

Hunninghake GW, Costabel U, Ando M, Baughman R, Cordier JF, du Bois R, Eklund A, Kitaichi M, Lynch J, Rizzato G, Rose C, Selroos O, Semenzato G, Sharma OP (1999) ATS/ERS WASOG statement on sarcoidosis. American Thoracic Society/European Respiratory Society/World Association of Sarcoidosis and other Granulomatous Disorders. Sarcoidosis Vasc Diffuse Lung Dis 16:149–173

Jee B, Katoch VM, Awasthi SK (2008) Dissection of relationship between small heat shock proteins and mycobacterial diseases. Indian J Lepr 80:231–245

Kalinowski L, Dobrucki LW, Brovkovych V, Malinski T (2002) Increased nitric oxide bioavailability in endothelial cells contributes to the pleiotropic effect of cerivastatin. Circulation 105:933–938

Klotz LO, Schroeder P, Sies H (2002) Peroxynitrite signaling: receptor tyrosine kinases and activation of stress-responsive pathways. Free Radic Biol Med 33:737–743

Liaudet L, Vassalli G, Pacher P (2009) Role of peroxynitrite in the redox regulation of cell signal transduction pathways. Front Biosci 14:4809–4814

Maertzdorf J, Weiner J 3rd, Mollenkopf HJ, Bornot T, Network TB, Bauer T, Prasse A, Müller-Quernheim J, Kaufmann SH (2012) Common patterns and disease-related signatures in tuberculosis and sarcoidosis. Proc Natl Acad Sci U S A 109:7853–7858

Magrì D, Mariotta S, Banfi C, Ricotta A, Onofri A, Ricci A, Pisani L, Cauti FM, Ghilardi S, Agostoni P (2013) Opposite behavior of plasma levels surfactant protein type B and receptor for advanced glycation end products in pulmonary sarcoidosis. Respir Med 107:1617–1624

Matzinger P (1994) Tolerance, danger, and the extended family. Ann Rev Immunol 12:991–1004

Mulvey MR, Switala J, Borys A, Loewen PC (1990) Regulation of transcription of katE and katF in Escherichia coli. J Bacteriol 172:6713–6720

Negri C, Scovassi AI, Cerino A, Negroni M, Borzì RM, Meliconi R, Facchini A, Montecucco CM, Astaldi Ricotti GC (1990) Autoantibodies to poly(ADP-ribose)polymerase in autoimmune diseases. Autoimmunity 6:203–209

Niforou K, Cheimonidou C, Trougakos IP (2014) Molecular chaperones and proteostasis regulation during redox imbalance. Redox Biol 30:323–332

Oswald-Richter KA, Beachboard DC, Zhan X, Gaskill CF, Abraham S, Jenkins C, Culver DA, Drake W (2010) Multiple mycobacterial antigens are targets of the adaptive immune response in pulmonary sarcoidosis. Respir Res 11:161

Peltoniemi M, Kaarteenaho-Wiik R, Säily M, Sormunen R, Pääkkö P, Holmgren A, Soini Y, Kinnula VL (2004) Expression of glutaredoxin is highly cell specific in human lung and is decreased by transforming growth factor-beta in vitro and in interstitial lung diseases in vivo. Hum Pathol 35:1000–1007

Piotrowski WJ, Górski P, Pietras T, Fendler W, Szemraj J (2011) The selected genetic polymorphisms of metalloproteinases MMP2, 7, 9 and MMP inhibitor TIMP2 in sarcoidosis. Med Sci Monit 17:598–607

Prabkhakanar K, Harris EB, Randhawa B (2000) Regulation by protein kinase of phagocytosis of *Mycobacterium leprae* by macrophages. J Med Microbiol 49:339–342

Szabo C (2003) Multiple pathways of peroxynitrite cytotoxicity. Toxicol Lett 140–141:105–112

Szabo C, Ischiropoulos H, Radi R (2007) Peroxynitrite: biochemistry, pathophysiology and development of therapeutics. Nat Rev Drug Discov 6:662–680

Thiagarajan G, Lakshmanan J, Chalasani M, Balasubramanian D (2004) Peroxynitrite reaction with eye lens proteins: α-Crystallin retains its activity despite modification. IOVS 45:2115–2121

Typiak MJ, Rębała K, Dudziak M, Dubaniewicz A (2014) Polymorphism of FCGR3A gene in sarcoidosis. Hum Immunol 75:283–288

van Maarsseveen TCM, Vos W, van Diest PJ (2009) Giant cell formation in sarcoidosis: cell fusion or proliferation with non-division? Clin Exp Immunol 155:476–486

Verdon CP, Burton BA, Prior RL (1995) Sample pretreatment with nitrate reductase and glucose-6-phosphate dehydrogenase quantitatively reduces nitrate while avoiding interference by $NADP^+$ when the Griess reaction is used to assay for nitrite. Anal Biochem 224:502–508

Voskuil MI, Schnappinger D, Visconti KC, Harrell MI, Dolganov GM, Herman DR, Schoolnik GK (2003) Inhibition of respiration by nitric oxide induces a *Mycobacterium tuberculosis* dormancy program. J Exp Med 198:705–713

Advs Exp. Medicine, Biology - Neuroscience and Respiration (2015) 15: 51–59
DOI 10.1007/5584_2015_138
© Springer International Publishing Switzerland 2015
Published online: 28 May 2015

Effects on Lung Function of Small-Volume Conventional Ventilation and High-Frequency Oscillatory Ventilation in a Model of Meconium Aspiration Syndrome

L. Tomcikova Mikusiakova, H. Pistekova, P. Kosutova, P. Mikolka, A. Calkovska, and D. Mokra

Abstract

For treatment of severe neonatal meconium aspiration syndrome (MAS), lung-protective mechanical ventilation is essential. This study compared short-term effects of small-volume conventional mechanical ventilation and high-frequency oscillatory ventilation on lung function in experimentally-induced MAS. In conventionally-ventilated rabbits, MAS was induced by intratracheal instillation of meconium suspension (4 ml/kg, 25 mg/ml). Then, animals were ventilated conventionally with small-volume (f-50/min; V_T-6 ml/kg) or with high frequency ventilation (f-10/s) for 4 h, with the evaluation of blood gases, ventilatory pressures, and pulmonary shunts. After sacrifice, left lung was saline-lavaged and cells in bronchoalveolar lavage fluid (BALF) were determined. Right lung was used for the estimation of lung edema formation (wet/dry weight ratio). Thiobarbituric acid-reactive substances (TBARS), oxidative damage markers, were detected in lung tissue and plasma. Meconium instillation worsened gas exchange, and induced inflammation and lung edema. Within 4 h of ventilation, high frequency ventilation improved arterial pH and CO_2 elimination compared with conventional ventilation. However, no other significant differences in oxygenation, ventilatory pressures, shunts, BALF cell counts, TBARS concentrations, or edema formation were observed between the two kinds of ventilation. We conclude that high frequency ventilation has only a slight advantage over small-volume conventional ventilation in the model of meconium aspiration syndrome in that it improves CO_2 elimination.

L. Tomcikova Mikusiakova, H. Pistekova, P. Kosutova,
P. Mikolka, A. Calkovska, and D. Mokra (✉)
Department of Physiology, Jessenius Faculty of Medicine
in Martin, Comenius University in Bratislava, 4 Mala
Hora St., SK-03601 Martin, Slovakia
e-mail: mokra@jfmed.uniba.sk

Keywords
Animal model • Aspiration syndrome • Lung injury • Meconium • Respiratory support • Ventilation

1   Introduction

Meconium aspiration syndrome (MAS) is defined as a respiratory distress in an infant born through the meconium-stained amniotic fluid with characteristic radiological changes, whose symptoms cannot be otherwise explained (Stenson and Smith 2012). Meconium is a sticky dark-green substance containing gastrointestinal secretions, bile and bile acids, mucus, pancreatic juice, blood, swallowed vernix caseosa, lanugo, and cellular debris. MAS is caused by aspiration of meconium in the airways during intrauterine gasping or during the first few breaths (van Ierland and de Beaufort 2009). MAS is present in 8–20 % of all deliveries, increasing to 23–52 % after 42nd week of gestation, whereas about 2–9 % of infants born through meconium fluid develop MAS (Swarnam et al. 2012). The respiratory manifestations include respiratory distress, tachypnea, cyanosis, end-expiratory grunting, alar flaring, and retractions of intercostal spaces. Barrel chest (increased anteroposterior diameter) due to the presence of air trapping can be observed and rales and rhonchi can be auscultated. Mortality rates for infants with MAS remain around 2.5–5.0 % (Stenson and Smith 2012).

The pathophysiology of MAS is complex. Aspirated meconium can interfere with breathing by several mechanisms. The pathophysiological mechanisms of hypoxemia in MAS include: (a) acute airway obstruction, (b) surfactant dysfunction, (c) chemical pneumonitis with release of vasoconstrictive and inflammatory mediators, (d) lung edema, and (e) persistent pulmonary hypertension of the newborn with right-to-left extrapulmonary shunting (Swarnam et al. 2012). An early phase ($<$15 min after meconium aspiration), primary caused by large airway obstruction, is characterized by an increase in lung resistance and functional residual capacity, a decrease in lung compliance and by hypoxemia, hypercarbia, and acidosis. Late phase ($>$60 min after meconium aspiration) results from spreading meconium distally, with subsequent obstruction of medium and small airways. In this phase, inflammatory changes and collapse of the airways and alveoli may be detected as a result of pneumonia and dysfunction of pulmonary surfactant (Mokra and Mokry 2010).

Supplemental oxygen administration is the mainstay of treatment for MAS and in less severe cases is the only therapy required (Singh et al. 2009). Of infants requiring mechanical respiratory support because of MAS, approximately 10–20 % is treated with continuous positive airway pressure alone. Approximately one-third of all infants with diagnosis of MAS require intubation and mechanical ventilation. Indications for intubation of infants with MAS include (a) high oxygen requirement ($FiO_2 > 0.8$), (b) respiratory acidosis with arterial pH persistently less than 7.25, (c) pulmonary hypertension, and (d) circulatory compromise with poor systemic blood pressure and perfusion (Dargaville 2012).

Despite more than four decades of mechanical ventilation for infants with MAS, the ventilatory management of MAS remains largely in the realm of 'art' rather than science, with very few clinical trials upon which to base definitive recommendations (Dargaville 2012). First described in the 1970s, high frequency oscillatory ventilation is a form of mechanical ventilation that uses small tidal volumes, sometimes less than anatomic dead space, and very rapid ventilator rates (2–20 Hz or 120–1,200 breaths/min). Potential advantages of this technique over conventional mechanical ventilation include the use of lower proximal airway pressures, the ability to adequately and independently manage oxygenation and ventilation while using extremely small tidal volumes, and the preservation of normal lung architecture, even when using high mean airway pressures (Goldsmith and Karotkin 2011). Despite a dearth of clinical trials or evidence suggesting a benefit, high frequency ventilation has become an important means of providing respiratory support for infants with severe MAS when conventional ventilation fails (Dargaville 2012). However, up to now no prospective randomized trials have clearly shown advantages of high frequency ventilation over conventional ventilation in MAS. Therefore, the aim of our study was to evaluate short-term effects of these two ventilatory modes on lung function in experimentally-induced MAS.

2 Methods

2.1 General Design of Experiments

Experimental protocols were performed in accordance with the ethical guidelines and authorized by the local Ethics Committee of Jessenius Faculty of Medicine in Martin, Comenius University in Bratislava and by the National Veterinary Board of Slovakia. Meconium was collected from healthy term neonates, lyophilized and stored at -20 °C. Before use, meconium was suspended in 0.9 % NaCl at a concentration of 25 mg/ml.

In the study, adult New Zealand white rabbits of both genders and mean body weight of 2.5 ± 0.3 kg were used. Animals were anesthetized with intramuscular ketamine (20 mg/kg; Narketan, Vétoquinol, UK) and xylazine (5 mg/kg; Xylariem, Riemser, Germany), followed by infusion of ketamine (20 mg/kg/h). Tracheotomy was performed and catheters were inserted into the femoral artery and right atrium for sampling the blood, and into the femoral vein to administer anesthetics. Animals were paralyzed with pipecuronium bromide (0.3 mg/kg/30 min; Arduan, Gedeon Richter, Hungary) and subjected to a neonatal

ventilator (SLE5000, SLE Limited, UK) and were ventilated conventionally with following settings: frequency (f) of 40/min, fraction of inspired oxygen (FiO$_2$) of 50 %, time of inspiration (Ti) of 50 %, peak inspiratory pressure (PIP)/ positive end-expiratory pressure (PEEP) of 1.0/ 0.4 kPa, and tidal volume (V$_T$) of 6 ml/kg. After 15 min of stabilization, respiratory parameters were recorded, and blood samples for analysis of blood gases (RapidLab 348, Siemens, Germany) and estimation of total and differential white blood cell (WBC) counts were taken.

Lung injury was induced by intratracheal administration of 4 ml/kg of meconium suspension (25 mg/ml), which was instilled into the endotracheal cannula in the semi-upright right and left lateral positions of the animal to provide homogenous lung distribution. When meconium-induced respiratory insufficiency developed, animals were randomly divided into two groups: with small-volume conventional mechanical ventilation (CMV group; f. 50/min, FiO$_2$ 100 %, Ti 50 %, PIP/PEEP 2.2/0.5 kPa, V$_T$ 6 ml/kg, mean airway pressure (MAP) 1.0 kPa) or high-frequency oscillatory ventilation (HFO group; f. 10/s, FiO$_2$ 100 %, MAP 1.0 kPa), and were ventilated with these ventilation settings for an additional 4 h. In high frequency ventilation, delta pressure was gradually decreasing from initial 1.8–2.0 kPa after MAS according to the actual value of PaCO$_2$, which we tried to keep in a range of 40–50 mmHg (5.33–6.67 kPa). In case of severe acidosis (pH < 7.2), animals were intravenously given sodium bicarbonate. Blood gases, WBC counts, and respiratory parameters were measured at 0.5, 1, 2, 3, and 4 h of the ventilatory treatment. At the end of experiment, animals were sacrificed by an overdose of anesthetics.

2.2 Measurement of Respiratory Parameters

In conventional ventilation, breathing rate, ventilatory pressures (PIP, PEEP, and MAP), V$_T$, FiO$_2$, minute ventilation (Vmin) and Ti; and in high frequency ventilation, breathing rate, Ti, delta pressure, MAP, V_T, Vmin, and FiO_2 were automatically measured by in-build sensors and software, and were displayed on the screen of the ventilator SLE5000. Oxygenation index (OI) was calculated as: $OI = (MAP \times FiO_2)/PaO_2$.

Right-to-left pulmonary shunts were calculated by a computer program using the Fick equation: $(CcO_2 - CaO_2)/(CcO_2 - CvO_2) \times 100$, where CcO_2, CaO_2, and CvO_2 are concentrations of oxygen in pulmonary capillaries, arterial, and mixed blood, respectively. CcO_2 was calculated by using P_AO_2 (alveolar partial pressure of oxygen) from the equation: $P_AO_2 = (PB - PH_2O) \times (FiO_2 - PaCO_2) \times [FiO_2 + (1 - FiO_2)/R]$, where PB is barometric pressure and PH_2O the pressure of water vapor. Respiratory exchange ratio (R) was assumed to be 0.8 and the current value of hemoglobin necessary for calculating the oxygen concentration in the blood was measured by combined analyzer (RapidLab 348, Siemens, Germany).

2.3 Counting of Cells in Bronchoalveolar Lavage Fluid and Arterial Blood

Samples of arterial blood for counting WBC were taken before creating of MAS model (initial) and at 0.5, 1, 2, 3, and 4 h of the treatment. Total WBC count was determined microscopically in a counting chamber after staining by Türck. Differential WBC count was estimated microscopically after panchromatic staining by May-Grünwald/Giemsa-Romanowski.

After sacrificing the animal, lungs, and trachea were excised. Left lung was lavaged with saline (0.9 % NaCl, 37 °C) 3×10 ml/kg, bronchoalveolar lavage fluid (BALF) was centrifuged at 1,500 rpm for 10 min. Total number of cells in BALF was determined microscopically in a counting chamber. Differential count of cells in the BALF sediment was evaluated microscopically after staining by May-Grünwald/Giemsa-Romanowski.

2.4 Lung Edema Formation (Wet/Dry Lung Weight Ratio)

Strips of the right lung tissue were cut, weighed and dried at 60 °C for 24 h to determine the wet/dry lung weight ratio.

2.5 Biochemical Analysis

Samples of arterial blood taken at the beginning of the experiment (initial), and at 2 and 4 h of ventilation therapy, were centrifuged (3,000 rpm, 15 min, 4 °C) and plasma was stored at -70 °C until the analysis was performed. Samples of right lung tissue were taken and prepared for additional biochemical analysis.

2.5.1 Preparation of the Lung Tissue Homogenate

Lung tissue was homogenized (5-times for 25 s, 1,200 rpm) in an ice-cold phosphate buffer (pH 7.4). Homogenates were then 3-times freezed and centrifuged (12,000 rpm, 15 min, 4 °C). Final supernatants were then stored at -70 °C until the analysis was performed. Protein concentrations in the lung homogenates were determined according to method described by Lowry et al. (1951), using bovine serum albumin as a standard.

2.5.2 Measurement of Thiobarbituric Acid-Reactive Substances (TBARS)

Formation of TBARS in the lung homogenate and plasma was measured by enzyme-linked immunosorbent ELISA method (OxiSelect™ TBARS Assay Kit; Cell Biolabs Inc., USA) and determined from the absorbance at 532 nm according to the manufacturer instructions, and results were expressed in µM.

2.6 Statistical Elaboration

Data were expressed as means $\pm$ SE. Between-group differences were analyzed by the Kruskal-Wallis test, within-group differences were evaluated by the Wilcoxon test. A value of P $<$ 0.05 was considered statistically significant. For analysis of data, a SYSTAT packet for Windows was used.

3   Results

Body weight of animals and values of the parameters before and 30 min after induction of the MAS model were comparable between the groups (all P $>$ 0.05).

3.1   Cellular Content in Bronchoalveolar Fluid and in Arterial Blood

Microscopic analysis of BALF showed slightly higher count of total cells (Fig. 1a) and neutrophils (Fig. 1b) in the animals with conventional ventilation; the differences between the CMV and HFO groups were insignificant (P $>$ 0.05). In the arterial blood, total count of circulating WBC was lower in CMV than that in HFO group, with a significant difference at 3 h of ventilation therapy (Fig. 2a.). Conversely to neutrophils in BALF, their count in blood was insignificantly lower in CMV than that in HFO group (P $>$ 0.05; Fig. 2b).

3.2   Biochemical Analysis

Within the first hours of creating MAS, concentrations of TBARS, the marker of oxidative injury, gradually increased in the plasma in both groups of animals, with no significant intergroup differences. Nor was there any significant difference in the concentration of TBARS in lung tissue homogenate (Table 1).

3.3   Lung Edema Formation

Ratio of wet and dry weight of the lung tissue expressing lung edema formation was comparable between the groups (P $>$ 0.05; Fig. 3).

3.4   Respiratory Parameters

Before and 30 min after creating the model of MAS, animals of both groups were ventilated with the corresponding values of ventilatory parameters: f-40/min, FiO_2-50 %, Ti-50 %, PIP/PEEP-1.0/0.4 kPa, and V_T-6 ml/kg. After induction of MAS, FiO_2 in both groups was increased to 100 % and MAP to approximately 1.0 kPa. During a 4-h period of ventilation, high frequency ventilation decreased $PaCO_2$ and improved arterial pH more effectively than

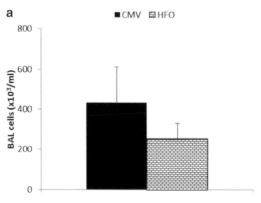
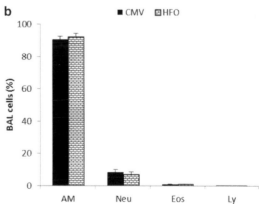

Fig. 1 (**a**) Total count of cells in the BALF at the end of experiment; (**b**) Differential count (percentage) of cells in the BALF at the end of experiment. *CMV* classical mechanical ventilation, *HFO* high frequency ventilation, *AM* alveolar macrophages, *Neu* neutrophils, *Eos* eosinophils, *Ly* lymphocytes. P $>$ 0.05 for between-group differences

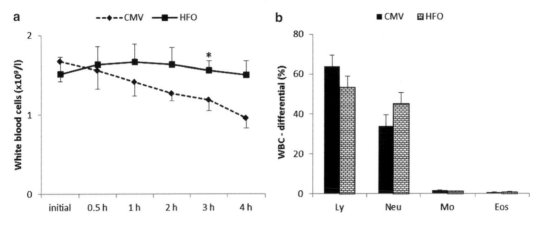

Fig. 2 (**a**) Total WBC count in the arterial blood during experiment; (**b**) Differential WBC count (percentage) in the arterial blood at the end of experiment. *CMV* classical mechanical ventilation, *HFO* high frequency oscillatory ventilation, *Ly* lymphocytes, *Neu* neutrophils, *Mo* monocytes, *Eos* eosinophils. *P < 0.05 for between-group difference

Table 1 Concentration of thiobarbituric acid-reactive substances (TBARS) in plasma (at the beginning of experiment, and at 2 and 4 h) and in lung homogenates at the end of experiment (4 h) in the CMV-ventilated and HFO-ventilated animals group

	TBARS plasma (μM)			TBARS lung (μM)
Time	Initial	2 h	4 h	4 h
CMV	10.79 ± 1.19	11.94 ± 1.50	15.50 ± 3.75	21.89 ± 2.23
HFO	17.65 ± 3.51	20.67 ± 4.51	20.93 ± 4.22	24.49 ± 3.00

CMV classical mechanical ventilation, *HFO* high frequency oscillatory ventilation
P > 0.05 for between-group comparisons

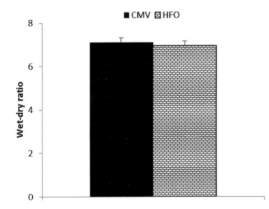

Fig. 3 Lung edema formation expressed as wet-dry lung weight ratio. P > 0.05 for between-group difference. *CMV* classical mechanical ventilation, *HFO* high frequency oscillatory ventilation

conventional ventilation (P < 0.05). In high frequency ventilation, there was also a trend for better oxygenation, but the improvement was statistically insignificant (Table 2).

4   Discussion

This study compared short-term effects of small-volume conventional mechanical ventilation and high-frequency oscillation on lung function in experimentally-induced meconium aspiration syndrome. Despite improved elimination of CO_2 and higher arterial pH in high frequency-ventilated animals, no significant changes in oxygenation, right-to-left pulmonary shunts, lung edema formation, or TBARS concentration in plasma and lung homogenate were found between the two modes of ventilation.

High frequency ventilation may improve blood gases because of other mechanisms of gas exchange which become active at high frequencies. High frequency ventilation reduces tidal volume needed to maintain ventilation, improves the uniformity of lung inflation, reduces air leak and intrapulmonary shunt, and

Table 2 Respiratory parameters before and after creating MAS and 0.5, 1, 2, 3, and 4 h of ventilation in the CMV-ventilated and HFO-ventilated animals

	Before MAS	After MAS	0.5 h	1 h	2 h	3 h	4 h
PaO$_2$ (kPa)							
CMV	32.5 ± 1.9	7.0 ± 0.6	9.4 ± 0.8	9.9 ± 0.9	10.5 ± 1.0	11.0 ± 1.2	10.6 ± 1.3
HFO	27.7 ± 1.8	7.8 ± 1.4	12.2 ± 4.4	12.0 ± 3.7	13.5 ± 3.6	15.6 ± 3.7	16.8 ± 4.6
OI							
CMV	0.8 ± 0.1	16.0 ± 2.1	13.1 ± 2.0	11.1 ± 1.1	10.4 ± 1.1	10.1 ± 1.2	10.8 ± 1.3
HFO	0.9 ± 0.1	13.8 ± 2.0	12.0 ± 2.2	13.7 ± 2.8	10.6 ± 2.3	9.2 ± 2.1	8.9 ± 2.1
PaCO$_2$ (kPa)							
CMV	5.8 ± 0.4	10.1 ± 0.8	9.2 ± 1.1	8.2 ± 0.9	7.3 ± 0.9	7.1 ± 0.6	6.7 ± 0.7
HFO	6.1 ± 0.4	8.6 ± 1.0	6.3 ± 0.4*	6.0 ± 0.3*	5.7 ± 0.2*	5.5 ± 0.2*	5.4 ± 0.2
pH$_a$							
CMV	7.40 ± 0.03	7.15 ± 0.02	7.12 ± 0.03	7.14 ± 0.02	7.11 ± 0.03	7.07 ± 0.03	7.03 ± 0.05
HFO	7.37 ± 0.03	7.23 ± 0.04	7.23 ± 0.01*	7.22 ± 0.02*	7.17 ± 0.03	7.13 ± 0.03	7.09 ± 0.03
RLS (%)							
CMV	5.8 ± 0.3	52.8 ± 1.5	55.4 ± 2.0	54.8 ± 1.9	53.1 ± 2.1	48.1 ± 2.4	41.3 ± 3.5
HFO	6.1 ± 0.5	49.4 ± 2.2	49.7 ± 3.4	51.4 ± 2.4	51.0 ± 2.7	48.8 ± 3.1	44.4 ± 2.9

CMV classical mechanical ventilation, *HFO* high frequency oscillatory ventilation, *PaO$_2$* partial pressure of oxygen in arterial blood, *OI* oxygenation index, *PaCO$_2$* partial pressure of carbon dioxide in arterial blood, *pH$_a$* pH in arterial blood, *RLS* right-to-left pulmonary shunts
*P < 0.05 for inter-group comparisons

improves oxygenation (Duval et al. 2009). The need for supplemental oxygen is reduced and exposure to free radicals are decreased.

In several animal studies, high frequency ventilation was superior to conventional ventilation in the short-term physiology and pressure exposure, and in the lung pathology over days to weeks. High frequency ventilation work at lower proximal airway pressures than conventional ventilation, reduced ventilator-induced lung injury and lung inflammatory markers, improved gas exchange, and decreased oxygen exposure (Froese et al. 1993; Jackson et al. 1994; Yoder et al. 2000). In a rabbit model of acute lung injury, high frequency ventilation showed better oxygenation, less histopathological injury score, and lower lung inflammatory responses, and oxidative stress in comparison with lung-protective conventional ventilation with V$_T$ of 6 ml/kg (Ronchi et al. 2011). In a study of Imai et al. (1994), high frequency ventilation decreased production of inflammatory mediators and resulted in less severe lung injury than conventional ventilation. In our present study, high frequency ventilation significantly decreased

arterial PCO$_2$ and pH while having airway pressures set comparably to those in conventional ventilation. There was also a trend to improved oxygenation, although the difference with conventional ventilation was not significant. In high frequency-ventilated group, a slightly lower count of BALF cells was found, associated with a higher count of circulating white blood cells, but there were no inter-group differences in lung edema formation or concentrations of TBARS in lung homogenate or in the plasma. Similarly to our findings, Renesme et al. (2013) found no apparent differences between the high frequency- and conventionally-ventilated meconium-instilled animals in the histologically evaluated lung injury.

Favorable pulmonary outcomes concerning high frequency ventilation observed in animal studies have not been consistently reproduced in human studies comparing the effects with those of conventional ventilation, either when looking at high frequency ventilation as an initial, elective mode of ventilation or as rescue ventilation when conventional ventilation failed to provide adequate gas exchange (Moriette et al. 2001;

Courtney et al. 2002; Johnson et al. 2002; Craft et al. 2003; Rojas et al. 2005). There are, however, some studies which showed some advantages of high frequency ventilation also in humans. For instance, Cools et al. (2010) showed high frequency ventilation to be equally effective to conventional ventilation in the preterm infants. In another study, Sun et al. (2014) showed that initial ventilation with high frequency ventilation in preterm infants with severe respiratory distress syndrome reduced the incidence of death and borderline personality disorder, and improved long-term neurodevelopment outcomes. In neonates with acute lung injury, rescue high frequency ventilation significantly improved oxygenation index, alveolar-arterial oxygen gradient, pH, PCO_2, and PO_2 and caused a better lung recruitment within 2 h (Poddutoor et al. 2011). In neonates treated for acute respiratory failure, significant decreases in mean airway pressure, FiO_2, and oxygenation index were found after starting high frequency ventilation, and $PaCO_2$ decreased after a further hour of ventilation (Jaballah et al. 2006).

In conclusion, high frequency ventilation improved elimination of CO_2 and supplied better arterial pH in comparison to conventional ventilation in this study. However, there were no other significant differences in oxygenation, right-to-left pulmonary shunts, edema formation, or lipid peroxidation between the animals ventilated with high frequency ventilation and small-volume conventional ventilation. The results indicate that both types of the lung-protective ventilation might be suitable for ventilation in meconium aspiration syndrome, but there is still a need for further clinical studies to determine optimal treatment strategies.

Acknowledgements Authors thank D. Kuliskova, Z. Remisova, M. Petraskova, and M. Hutko for technical assistance. In addition, we would like to thank for support to projects APVV-0435-11, VEGA 1/0305/14, and BioMed (ITMS 26220220187).

Conflicts of Interest The authors declare no conflict of interest in relation to this article.

References

Cools F, Askie LM, Offringa M, Asselin JM, Calvert SA, Courtney SE, Dani C, Durand DJ, Gerstmann DR, Henderson-Smart DJ, Marlow N, Peacock JL, Pillow JJ, Soll RF, Thome UH, Truffert P, Schreiber MD, Van Reempts P, Vendettuoli V, Vento G, PreVILIG collaboration (2010) Elective high-frequency oscillatory versus conventional ventilation in preterm infants: a systematic review and meta-analysis of individual patients' data. Lancet 375:2082–2091

Courtney SE, Durand DJ, Asselin JM, Hudak ML, Aschner JL, Shoemaker CT, Neonatal Ventilation Study Group (2002) High-frequency oscillatory ventilation versus conventional mechanical ventilation for very-low-birth-weight infants. N Engl J Med 347:643–652

Craft AP, Bhandari V, Finer NN (2003) The sy-fi study: a randomized prospective trial of synchronized intermittent mandatory ventilation versus a high-frequency flow interrupter in infants less than 1000 g. J Perinatol 23:14–19

Dargaville PA (2012) Respiratory support in meconium aspiration syndrome: a practical guide. Int J Pediatr 2012:965159. doi:10.1155/2012/965159

Duval ELIM, Markhorst DG, van Vught AJ (2009) High-frequency oscillatory ventilation in children: an overview. Respir Med CME 2:155–161

Froese AB, McCulloch PR, Sugiura M, Vaclavik S, Possmayer F, Moller F (1993) Optimizing alveolar expansion prolongs the effectiveness of exogenous surfactant therapy in the adult rabbit. Am Rev Respir Dis 148:569–577

Goldsmith JP, Karotkin EH (eds) (2011) Assisted ventilation of the neonate, 3rd edn. W.B. Saunders, St. Louis, p 656

Imai Y, Kawano T, Miyasaka K, Takata M, Imai T, Okuyama K (1994) Inflammatory chemical mediators during conventional ventilation and during high frequency oscillatory ventilation. Am J Respir Crit Care Med 150:1550–1554

Jaballah N, Khaldi A, Mnif K, Bouziri A, Belhadj S, Hamdi A, Kchaou W (2006) High-frequency oscillatory ventilation in pediatric patients with acute respiratory failure. Pediatr Crit Care Med 7:362–367

Jackson JC, Truog WE, Standaert TA, Murphy JH, Juul SE, Chi EY, Hildebrandt J, Hodson WA (1994) Reduction in lung injury after combined surfactant and high-frequency ventilation. Am J Respir Crit Care Med 150:534–539

Johnson AH, Peacock JL, Greenough A, Marlow N, Limb ES, Marston L, Calvert SA, United Kingdom Oscillation Study Group (2002) High-frequency oscillatory ventilation for the prevention of chronic lung disease of prematurity. N Engl J Med 347:633–642

Lowry OH, Rosebrough NJ, Farr AL, Randall RJ (1951) Protein measurement with the Folin phenol reagent. J Biol Chem 193:265–275

Mokra D, Mokry J (2010) Meconium aspiration syndrome: from pathomechanisms to treatment, 1st edn. Nova Science Pub Inc., New York

Moriette G, Paris-Llado J, Walti H, Escande B, Magny JF, Cambonie G, Thiriez G, Cantagrel S, Lacaze-Masmonteil T, Storme L, Blanc T, Liet JM, André C, Salanave B, Bréart G (2001) Prospective randomized multicenter comparison of high-frequency oscillatory ventilation and conventional ventilation in preterm infants of less than 30 weeks with respiratory distress syndrome. Pediatrics 107:363–372

Poddutoor PK, Chirla DK, Sachane K, Shaik FA, Venkatlakshmi A (2011) Rescue high frequency oscillation in neonates with acute respiratory failure. Indian Pediatr 48:467–470

Renesme L, Elleau C, Nolent P, Fayon M, Marthan R, Dumas De la Roque E (2013) Effect of high-frequency oscillation and percussion versus conventional ventilation in a piglet model of meconium aspiration. Pediatr Pulmonol 48:257–264

Rojas MA, Lozano JM, Rojas MX, Bose CL, Rondón MA, Ruiz G, Piñeros JG, Rojas C, Robayo G, Hoyos A, Celis LA, Torres S, Correa J, Columbian Neonatal Research Network (2005) Randomized, multicenter trial of conventional ventilation versus high-frequency oscillatory ventilation for the early management of respiratory failure in term or near-term infants in Colombia. J Perinatol 25:720–724

Ronchi CF, dos Anjos Ferreira AL, Campos FJ, Kurokawa CS, Carpi MF, de Moraes MA, Bonatto RC, Defaveri J, Yeum KJ, Fioretto JR (2011) High-frequency oscillatory ventilation attenuates oxidative lung injury in a rabbit model of acute lung injury. Exp Biol Med (Maywood) 236:1188–1196

Singh BS, Clark RH, Powers RJ, Spitzer AR (2009) Meconium aspiration syndrome remains a significant problem in the NICU: outcomes and treatment patterns in term neonates admitted for intensive care during a ten-year period. J Perinatol 29:497–503

Stenson BJ, Smith CL (2012) Management of meconium aspiration syndrome. Paediatr Child Health 22:532–535

Sun H, Cheng R, Kang W, Xiong H, Zhou C, Zhang Y, Wang X, Zhu C (2014) High-frequency oscillatory ventilation versus synchronized intermittent mandatory ventilation plus pressure support in preterm infants with severe respiratory distress syndrome. Respir Care 59:159–169

Swarnam K, Soraisham AS, Sivanandan S (2012) Advances in the management of meconium aspiration syndrome. Int J Pediatr 2012:359571

van Ierland Y, de Beaufort AJ (2009) Why does meconium cause meconium aspiration syndrome? Current concepts of MAS pathophysiology. Early Hum Dev 85:617–620

Yoder BA, Siler-Khodr T, Winter VT, Coalson JJ (2000) High-frequency oscillatory ventilation: effects on lung function, mechanics, and airway cytokines in the immature baboon model for neonatal chronic lung disease. Am J Respir Crit Care Med 162:1867–1876

Advs Exp. Medicine, Biology - Neuroscience and Respiration (2015) 15: 61–69
DOI 10.1007/5584_2015_144
© Springer International Publishing Switzerland 2015
Published online: 29 May 2015

Expression of HIF-1A/VEGF/ING-4 Axis in Pulmonary Sarcoidosis

W.J. Piotrowski, J. Kiszałkiewicz, D. Pastuszak-Lewandoska,
P. Górski, A. Antczak, M. Migdalska-Sęk, W. Górski,
K.H. Czarnecka, D. Domańska, E. Nawrot,
and E. Brzeziańska-Lasota

Abstract

Angiogenesis/angiostasis regulated by hypoxia inducible factor-1A (HIF-1A)/vascular endothelial growth factor (VEGF)/inhibitor of growth protein 4 (ING-4) axis may be crucial for the course and outcome of sarcoidosis. Overexpression of angiogenic factors (activation of VEGF through HIF-1A) may predispose to chronic course and lung fibrosis, whereas immunoangiostasis (related to an overexpression of inhibitory ING-4) may be involved in granuloma formation in early sarcoid inflammation, or sustained or recurrent formation of granulomas. In this work we investigated gene expression of HIF-1A, VEGF and ING-4 in bronchoalveolar fluid (BALF) cells and in peripheral blood (PB) lymphocytes of sarcoidosis patients (n = 94), to better understand mechanisms of the disease and to search for its biomarkers. The relative gene expression level (RQ value) was analyzed by qPCR. The results were evaluated according to the presence of lung parenchymal involvement (radiological stage I *vs.* II–IV), acute *vs.* insidious onset, lung function tests, calcium metabolism parameters, percentage of lymphocytes (BALL %) and BAL $CD4^+/CD8^+$ in BALF, age, and gender. In BALF cells, the ING-4 and VEGF RQ values were increased, while HIF-1A expression was decreased. In PB lymphocytes all studied genes were overexpressed. Higher expression of HIF-1A in PB lymphocytes of patients with abnormal spirometry, and in BALF cells of patients with lung volume

W.J. Piotrowski, P. Górski, and W. Górski
Department of Pneumology and Allergy, Medical
University of Lodz, 251 Pomorska St., 92-213 Lodz,
Poland

J. Kiszałkiewicz, D. Pastuszak-Lewandoska,
M. Migdalska-Sęk, K.H. Czarnecka, D. Domańska,
E. Nawrot, and E. Brzeziańska-Lasota (✉)
Department of Molecular Bases of Medicine, Medical
University of Lodz, 251 Pomorska St., 92-213 Lodz,
Poland
e-mail: ewa.brzezianska@umed.lodz.pl

A. Antczak
Department of General and Oncological Pulmonology,
First Chair of Internal Diseases, Medical University of
Lodz, 251 Pomorska St., 92-213 Lodz, Poland

restriction was found. VEGF gene expression in BALF cells was also higher in patients with abnormal spirometry. These findings were in line with previous data on the role of HIF-1A/VEGF/ING-4 axis in the pathogenesis of sarcoidosis. Up-regulated HIF-1A and VEGF genes are linked to acknowledged negative prognostics.

Keywords

Biomarkers • Disease mechanisms • Lung fibrosis • Prognosis • Radiological classification

1   Introduction

Sarcoidosis is an inflammatory granulomatous disorder of unknown etiology, affecting multiple organs, but mainly intrathoracic lymph nodes and the lungs (Baughman et al. 2011; Hunninghake et al. 1999). It has been reported in all racial and ethnic groups. In the majority of patients the disease disappears without treatment, but it may become chronic and progressive, leading to debilitating lung fibrosis, and in some cases also to death (Ianuzzi et al. 2007). Genetic susceptibility seems to be crucial for the development of the disease in subjects exposed to unknown environmental (infectious or non-infectious) factors (antigens) (Lazarus 2009; Ianuzzi et al. 2007; Thomas and Hunninghake 2003). Genetic factors strongly influence the disease course and long-term prognosis. However, it is very difficult to predict in the clinical setting, which patients are at increased risk of unfavorable outcome in the future.

A few studies have been published recently, showing that angiogenic and angiostatic factors may be involved in the pathogenesis of sarcoidosis (Cui et al. 2010; Lazarus 2009; Ianuzzi et al. 2007; Antoniou et al. 2006). It is claimed that angiogenesis-angiostasis balance is implicated in the formation of granuloma (angiostasis), and in hypoxia-induced lung remodeling and fibrosis (angiogenesis) (Cui et al. 2010). Hypoxia-inducible factor-1A (HIF-1A, located in the chromosome 14q21-q24) is a key transcription factor in the cellular response to hypoxia, and is recognized as a major oxygen homeostasis regulator which controls the activation of genes essential to cellular adaptation to low oxygen conditions, such as vascular endothelial growth factor (VEGF). However, also under normoxic conditions, several dozen of genes implicated in different cellular functions have been found to be induced by growth factors and vascular hormones through the mediation of HIF-1A (Déry et al. 2005). HIF-1 is a key regulator of VEGF, which stimulates mobility and maturation of endothelial cells in hypoxic environment. In addition, the group of HIF-1A dependent genes (including MMP2) influence the extracellular matrix turnover. Under hypoxic conditions, HIF-1 upregulation may induce alveolar cell apoptosis and epithelial-mesenchymal transmission (EMT), thus potentially contributing to pulmonary fibrosis (Kim et al. 2006). VEGF, by controlling monocyte recruitment, may be involved in granuloma formation (Tzouvelekis et al. 2012; Tolnay et al. 1998). In chronic experimental hypoxia, leading to acute lung fibrosis, inflammation, fibrosis, and pulmonary hypertension, the increased HIF-1A protein and its DNA binding activity have been observed in pulmonary epithelial cells (Shimoda and Semenza 2011; Yu et al. 1998). Recently, it has been documented that HIF-1A dependent lung epithelial remodeling involves changes in the profile of chemokines and cytokines, properties of myofibroblasts, and a variety of proangiogenic factors, including VEGF (Stenmark et al. 2006). Inhibitor of growth protein 4 (ING-4) has been recognized as a potential tumor suppressor gene, and effective suppressor of HIF-1A. It is involved in the regulation of cell cycle arrest, apoptosis or senescence, inhibiting cell proliferation and angiogenesis (Guérillon et al. 2014).

Currently, no useful diagnostic or prognostic biomarkers are available to support the clinical judgment of patients suffering from sarcoidosis. Moreover, relatively little is known about the significance of HIF1-A/VEGF/ING-4 axis in sarcoidosis development and disease course. Therefore, in the present study we analyzed the expression levels of HIF-1A, VEGF, ING-4 genes and evaluated their potential diagnostic and prognostic value in sarcoidosis patients.

2 Methods

2.1 Study Group

The study was approved by Ethics Committee of the Medical University of Lodz, Poland (permission RNN/141/10/KE) and patients participating in the study signed written informed consent. This study is part of the project on research into the molecular mechanisms underlying the pathogenesis of sarcoidosis. The project encompassed 94 hospitalized patients with the diagnosis of pulmonary sarcoidosis and 50 non-smoking patients suffering from idiopathic cough or other unrelated to sarcoidosis conditions. All subject were recruited for the study in the years 2010–2014. The diagnosis was based on the current international guidelines (Ianuzzi et al. 2007; ATS & ERS & WASOG 1999). Clinical and radiological features of sarcoidosis, with the presence of non-caseating granuloma in tissue biopsy, were confirmed in each patient. Demographic characteristics of the sarcoidosis and control patients have been previously described in detail (Piotrowski et al. 2014a).

The current study on the role in sarcoid formation of angiogenesis/angiostasis balance controlled by HIF-1A/VEGF interplay expands on the previous investigation concerning the expression of the cytokine transforming growth factor beta (TGF-β)/SMAD receptor signaling cascade in the development of clinical features of sarcoidosis. The source of research material for the current study were the same samples of bronchoalveolar lavage fluid (BALF) and peripheral blood which were used in research on the TGF-β/SMAD cascade previously published (Piotrowski et al. 2014a). The angiogenesis/angiostasis status was considered an independent research ramification of the molecular pathogenesis of sarcoidosis, and, therefore, was herein described as a separate entity.

2.2 Collection of Biological Material

Bronchoscopy was performed with a flexible bronchoscope (Pentax, Tokyo, Japan) according to the Polish Respiratory Society Guidelines (Chciałowski et al. 2011). BALF was collected from the medial lung lobe, by instillation and subsequent withdrawal of 4×50 ml of 0.9 % NaCl. The fluid recovery was 52.1 ± 1.2 %. The crude BALF was filtered through a gauze to clear the thick mucus and other contaminants, centrifuged, and the pellet was suspended in a phosphate buffer. The total number of non-epithelial cells (total cell count – TCC) was presented as $n \times 10^6$. Cytospin slides were prepared and stained by May-Grünwald-Giemsa stain. The number of macrophages, lymphocytes, neutrophils, and eosinophils was calculated under a light microscope and presented as percent of TCC. After the calculations, all fluid was centrifuged (10 min, 1,200 rpm), supernatant of BALF was suspended in RNAlater RNA Stabilization Reagent (QIAGEN, Hilden, Germany) in a volume of about 350 µl of solution in Eppendorf tubes, marked with an identification number, and was frozen (-80 °C) until further RNA isolation procedures.

Spirometry was performed according to the Polish Respiratory Society Guidelines (Polish Society of Respiratory Diseases 2006) with a computer-based spirometer (Jaeger, Dortmund, Germany).

Blood was collected into 5 ml EDTA containing tubes. For lymphocyte separation, a density gradient cell separation medium Histopaque-1077 (Sigma-Aldrich, Poznan, Poland) was used according to the manufacturer's protocol.

2.3   RNA Extraction, Real-Time qPCR

RNA isolation was performed using mirVana™ miRNA Isolation Kit (Life Technologies, Carlsbad, CA), according to the manufacturer's protocol. The quality and quantity of isolated RNA was assessed spectrophotometrically (BioPhotometerTM Plus, Eppendorf, Hamburg, Germany). The purity of total RNA (ratio of 16S to 18S fraction) was determined in the automated electrophoresis using RNA Nano Chips LabChipplates on Agilent 2100 Bioanalyzer (Agilent Technologies, Santa Clara, CA).

cDNA was transcribed from 100 ng of total RNA, using a High-Capacity cDNA Reverse Transcription Kit (Applied Biosystems, Carlsbad, CA) in a total volume of 20 μl according to manufacturer's protocol. The relative expression analysis was performed in 7900HT fast real-time PCR System (Applied Biosystems, Carlsbad, CA) using TaqMan probes for the studied genes: *HIF-1A* (Hs00153153_m1), *VEGF* (Hs00900055_m1), *ING-4* (Hs01088026_m1), *ACTB* (Hs99999903_m1). The PCR mixture contained: cDNA (1 to 100 ng), 20 × TaqManR Gene Expression Assay, 2 × KAPA PROBE Master Mix (2x) ABI Prism Kit (Kapa Biosystems, Wilmington, MA) and RNase-free water in a total volume of 20 μl. The expression levels (RQ values) of the studied genes were calculated using the delta CT method, with the adjustment to the β-actin expression level and in relation to the expression level of calibrator (Human Lung Total RNA Ambion®; Applied Biosystems, Carlsbad, CA), for which RQ value was equal to 1.

2.4   Data Elaboration

The Kruskal-Wallis test, the Mann-Whitney U test, the Neuman–Keuls multiple comparison test, and the Spearman rank correlation were used for statistical data elaboration (StatSoft, Cracow, Poland). Statistically significant differences were defined as $P < 0.05$.

3   Results

3.1   Relative Expression Analysis of Genes in Bronchoalveolar Lavage Fluid Cells

In BALF cells from patients with radiological stage I sarcoidosis, the highest expression level (mean RQ) of *ING-4* (0.629), and the lowest of *HIF-1A* (0.037) were observed. Similarly, in patients with radiological stages II–IV, the highest RQ values were revealed for *ING-4* (0.634), and the lowest for *HIF-1A* (0.031).

In the acute onset phenotype of sarcoidosis, the highest expression level (mean RQ) was observed for *ING-4* (0.568) and the lowest for *HIF-1A* (0.033). Similarly, in the chronic insidious onset phenotype, the highest expression level (mean RQ) was found for *ING-4* (0.686), and the lowest for *HIF-1A* (0.035). Decreased (RQ < 1) and increased (RQ > 1) expression values of studied genes are shown in Table 1.

3.2   Relative Expression Analysis of Genes in Peripheral Blood Lymphocytes

In peripheral blood (PB) lymphocytes in patients with radiological stage I sarcoidosis, the highest mean expression level (mean RQ) of *ING-4* (0.540) and the lowest of *HIF-1A* (0.321) were observed. In patients with radiological stages II–IV, the highest mean expression level of *ING-4* (0.586) and the lowest of *VEGF* (0.025) were found.

In the acute onset phenotype, the highest mean expression level of *ING-4* (0.592) and the lowest of *HIF-1A* (0.246) were observed. Similarly, in the insidious onset phenotype, the highest mean expression level for *ING-4* (0.526) and the lowest for *VEGF* (0.089) were found. Decreased (RQ < 1) and increased (RQ > 1) expression values of studies genes are shown in Table 2.

Table 1 Mean RQ values (range) for all studied genes (*HIF-1A, ING-4, and VEGF*) in BALF cells of sarcoidosis patients

			Number (%) of samples with	
		Mean RQ value (range)	RQ value >1	RQ value <1
HIFI-A	Radiological stage I	0.037 (0.002–0.529)	0 (0)	46 (100)
	Radiological stages II–IV	0.031 (0.001–0.144)	0 (0)	48 (100)
	Acute disease	0.033 (0.002–0.529)	0 (0)	42 (100)
	Chronic disease	0.035 (0.001–0.453)	0 (0)	51 (100)
ING-4	Radiological stage I	0.629 (0.018–3.555)	10 (23)	36 (77)
	Radiological stages II–IV	0.634 (0.024–4.448)	11 (22)	37 (78)
	Acute disease	0.568 (0.330–3.555)	9 (22)	33 (78)
	Chronic disease	0.686 (0.024–4.448)	9 (18)	42 (82)
VEGF	Radiological stage I	0.250 (0.002–3.320)	5 (11)	41 (89)
	Radiological stages II–IV	0.098 (0.001–1.710)	2 (5)	46 (95)
	Acute disease	0.271 (0.002–3.320)	5 (12)	37 (88)
	Chronic disease	0.094 (0.001–1.717)	1 (2)	50 (98)

RQ gene expression level, *BALF* bronchoalveolar fluid

Table 2 Mean RQ value (range) for all studied genes (*HIF-1A, ING-4, and VEGF*) in PB lymphocytes in sarcoidosis patients

			Number (%) of samples with	
		Mean RQ value (range)	RQ value >1	RQ value <1
HIF-1A	Radiological stage I	0.321 (0.009–2.589)	2 (10)	19 (90)
	Radiological stages II–III	0.123 (0.007–0.950)	0 (0)	17 (100)
	Acute disease	0.246 (0.009–2.252)	1 (5)	19 (95)
	Chronic disease	0.207 (0.029–2.589)	1 (5)	17 (95)
ING-4	Radiological stage I	0.540 (0.013–2.994)	2 (10)	19 (90)
	Radiological stages II–III	0.586 (0.025–3.369)	3 (18)	14 (82)
	Acute disease	0.592 (0.013–2.994)	3 (15)	17 (75)
	Chronic disease	0.526 (0.025–3.361)	2 (11)	16 (89)
VEGF	Radiological stage I	0.336 (0.003–2.887)	3 (13)	19 (87)
	Radiological stages II–III	0.025 (0.005–0.178)	0 (0)	17 (100)
	Acute disease	0.295 (0.003–2.877)	2 (10)	18 (90)
	Chronic disease	0.089 (0.002–1.152)	1 (5)	17 (95)

RQ gene expression level, *PB* Peripheral blood

3.3 Expression of Genes in Sarcoidosis Patients *vs.* Controls

BALF Cells In BALF cells, in the whole group of sarcoidosis patients, the mean RQ values of the genes were the following: 0.030 for *HIF1-A*, 0.172 for *VEGF*, and 0.703 for *ING-4* gene (for specific clinical classification see Table 1). In the control group, the RQ values were the following: 0.050 for both *HIF1-A* and *VEGF*, and 0.331 for *ING-4* gene. There were no statistically

significant differences regarding the level of genes expression between the sarcoidosis and control groups (P > 0.05; Mann-Whitney U test).

PB Lymphocytes In PB lymphocytes, in the whole group of sarcoidosis patients, the mean RQ values of the genes were the following: 0.233 for *HIF1-A*, 0.197 for *VEGF*, and 0.561 for *ING-4* gene (for specific clinical classification see Table 2). In the control group, the RQ values were the following: 0.024 for *HIF-1A*, 0.013 for *VEGF*, and 0.468 for *ING-4* gene. The level of expression of the *HIF-1A* gene was here

Fig. 1 Expression levels of *HIF-1A* in bronchoalveolar lavage fluid (BALF) cells *vs.* peripheral blood (PB) lymphocytes in sarcoidosis patients

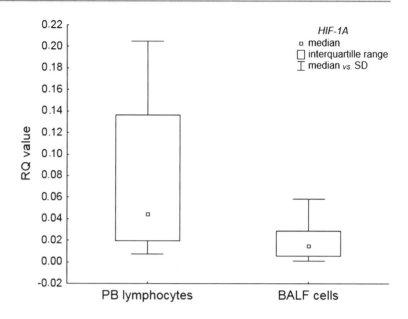

significantly greater in sarcoidosis patients than that in the control subjects (P = 0.003, Mann-Whitney U test).

3.4 Expression of Genes in BALF Cells *vs.* PB Lymphocytes in Sarcoidosis Patients

There was significantly greater *HIF-1A* expression in PB lymphocytes than that in BALF cells in sarcoidosis patients (P <0.0001; Mann-Whitney U test) (Fig. 1).

3.5 Expression of Genes in Relation to Radiological and Clinical Classification and Lung Function

In both BALF cells and PB lymphocytes, there were no significant differences in the expression of the genes studied between the subgroups with and without parenchymal involvement (stage I *vs.* II–IV), and between the acute *vs.* insidious onset phenotypes. However, a significantly higher expression level of *HIF-1A* in BALF cells in patients with abnormal spirometry, especially those with lung volume restriction pattern

was found (P = 0.033) (Fig. 2). We also observed a significantly higher expression level of *VEGF* gene in BALF cells in patients with abnormal spirometry (P = 0.028).

3.6 Relationship Between the Expression of Genes and Lung Function, Patients' Features, and Laboratory Markers in Bronchoalveolar Lavage Fluid Cells and in Peripheral Blood Lymphocytes

Several negative correlations were found between the *HIF-1A* and *VEGF* gene expression in BALF, and *VEGF* expression in PB lymphocytes, on one side, and spirometric tests or DLCOc on the other side (Table 3).

4 Discussion

In the present study we documented down-regulated *HIF-1A*, and up-regulated *VEGF* and *ING-4* gene expression in BALF cells, while up-regulated gene mRNAs of the entire axis was observed in peripheral blood lymphocytes. Our results are partly in accord with the observations

Fig. 2 Expression levels of *HIF-1A* gene in patients with normal spirometry and with lung volume restriction (P = 0.033; Mann-Whitney U test)

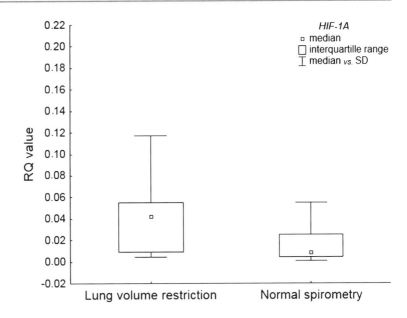

Table 3 Correlations between the expression level of genes (RQ values) and lung function tests and selected laboratory markers in sarcoidosis patients

Biological material	Parameter	Gene	Group/subgroup	Rho value	P value
BALF cells	FEV_1/FVC	*HIF-1A*	Entire	−0.217	0.048
	FEV_1/FVC	*HIF-1A*	Parenchymal involvement	−0.322	0.037
	DLCOc	*HIF-1A*	Entire	−0.387	0.002
	DLCOc	*VEGF*	Entire	−0.300	0.021
	CD4/CD8	*VEGF*	Without parenchymal involvement	−0.331	0.039
PB lymphocytes	FEV_1	*VEGF*	Parenchymal involvement	−0.650	0.006
	FVC	*VEGF*	Parenchymal involvement	−0.562	0.023
	DLCOc	*VEGF*	Parenchymal involvement	−0.612	0.001

BALF bronchoalveolar lavage fluid, *PB* peripheral blood, *DLCOc* lung diffusion capacity for carbon monoxide corrected for hemoglobin, FEV_1 forced expiratory volume in 1 s, *FVC* forced vital capacity

of other authors (Tzouvelekis et al. 2012) who also found suppressed *HIF-1A* and abundant *VEGF* and *ING-4* expression in tissue biopsies containing sarcoid granuloma. This unusual biological cascade of genes expression results, according to the cited authors, from the functional feedback between *HIF-1A* and *ING-4* genes. Interestingly, we also observed a negative interaction between *ING-4* and *HIF-1A*, as the overexpression of *ING-4* frequently coexisted with decreased expression of *HIF-1A* in BALF cells in sarcoidosis patients. Other authors also confirmed that *ING-4* is able to inhibit *HIF-1A*, or *ING-4* may also be suppressed by other growth factors present in the cytokine milieu of sarcoid

inflammatory site (Tzouvelekis et al. 2012; Antoniou et al. 2006; Déry et al. 2005).

The known mechanism of inhibitory properties of *ING-4* is based on the interaction with *HIF* prolyl hydroxylases and NF-κB subunit – ReIA. The net effect is the repression of angiogenesis related genes, such as *IL-6*, *IL-8*, and *COX-2*. Immunoangiostatic environment in sarcoidosis may also be confirmed by the studies showing the role of CXCR3 ligands (interferon inducible cytokines, such as IP-10, ITAC, and MIG). These cytokines are related to more severe and chronic course of sarcoidosis (Piotrowski et al. 2014b). Apparently, the inconsistent rise in *VEGF* gene expression is the evidence of axis impairment and may be explained by a dual character of sarcoid inflammation – on the one side immunoangiostatic and on the other side inflammatory, in which VEGF may also be involved in inflammatory pathways different than angiogenesis. However, this dual immunoangiostatic and inflammatory nature of sarcoid granuloma formation involves stimuli unrelated to HIF-1/VEGF axis (Tzouvelekis et al. 2012). An elevated concentration of VEGF protein in the serum and BALF of sarcoidosis patients and increased VEGF protein expression in tissue biopsies has also been found by Yamashita et al. (2013). Additionally, VEGF concentration has been shown to be higher in patients with extrapulmonary involvement, and the level of VEGF is helpful in predicting which patients would deserve corticosteroid treatment (Sekiya et al. 2003). A comparison of VEGF concentration in BALF retrieved from lung segments with more and less intensive parenchymal changes revealed a higher concentration of VEGF in more involved lung areas. Moreover, the relation to the disease activity was confirmed (Ziora et al. 2000). This up-regulation of *VEGF* on the translational level has also been confirmed in the present study on the mRNA level, and a negative correlation between *VEGF* expression, and lung function results and parenchymal involvement in sarcoidosis patients was also recognized. Our observation concerning the up-regulation of VEGF is important for the recognition of this gene as a strong angiogenic stimulator, which activates mobility of endothelial cells toward the hypoxic environment and in consequence leads to lung epithelial remodeling. Therefore, *VEGF* expression and immunoexpression in the serum or BALF seem to have a prognostic value in sarcoidosis patients.

In idiopathic pulmonary fibrosis (IPF), a chronic progressive disease leading to extensive interstitial fibrosis, the pro-angiogenic environment has been documented. That is interesting in the context of possible mechanisms of lung fibrosis in the course of sarcoidosis. In a study comparing sarcoidosis and IPF, the increased *VEGF* mRNA expression in BALF of IPF patients, and also lower expression of *CXCL12* and *CXCR4* mRNA, constitute the pro-angiogenic microenvironment (Antoniou et al. 2009). The same group of authors reported a different profile of pro-angiogenic (GRO-a, ENA-78, IL-8) and anti-angiogenic (CXCR3 ligands) cytokines in BALF, with increased levels of pro-angiogenic cytokines in IPF (Antoniou et al. 2006). This is in line with the observation of other authors who found a negative correlation between angiogenic activity of sera of patients with interstitial lung disease and their lung diffusion capacity (Zielonka et al. 2010).

In the present study, we also documented the negative correlation in PB lymphocytes between expression of mRNA of another angiogenic factor, i.e., *HIF-1A* and lung function (FEV_1/FVC) in sarcoidosis patients with parenchymal involvement. It seems an important observation due to the fact that up-regulation of *HIF-1A* on the translational level may inhibit the epithelial cell repair mechanisms, which facilitates the development of pulmonary fibrosis. The process of pulmonary fibrosis, as a consequence of granuloma healing in the lung sarcoidosis, may impair respiratory function and has to do with diseases progression. Therefore, the expression of *HIF-1A* mRNA may be useful as a progression marker in sarcoidosis. Reassuming, up-regulation of the pro-angiogenic *HIF-1A* and *VEGF* genes observed in the present study is linked to the acknowledged negative prognostic factors in sarcoidosis. However, the results cannot provide data confirming direct relationship between the shift to pro-angiogenic environment and progression of sarcoidosis toward extensive lung fibrosis. That would deserve a separate follow-up study, with the recruitment of a number of patients with disease progression.

Acknowledgements This work was partially funded by the grant of the National Science Center (grant no. 2011/01/B/NZ5/04239).

Conflicts of Interest The authors declare no conflicts of interest in relation to this article.

References

American Thoracic Society: European Respiratory Society: World Association of Sarcoidosis and Other Granulomatous Disorders (1999) Statement on sarcoidosis. Am J Respir Crit Care Med 160:736–755

Antoniou KM, Tzouvelekis A, Alexandrakis MG, Sfiridaki K, Tsiligianni I, Rachiotis G, Tzanakis N, Bouros D, Milic-Emili J, Siafakas NM (2006) Different angiogenic activity in pulmonary sarcoidosis and idiopathic pulmonary fibrosis. Chest 130:982–988

Antoniou KM, Soufla G, Proklou A, Margaritopoulos G, Choulaki C, Lymbouridou R, Samara KD, Spandidos DA, Siafakas NM (2009) Different activity of the biological axis VEGF-Flt-1 (fms-like tyrosine kinase 1) and CXC chemokines between pulmonary sarcoidosis and idiopathic pulmonary fibrosis: a bronchoalveolar lavage study. Clin Dev Immunol 537929. doi:10.1155/2009/537929

Baughman RP, Nagai S, Balter M, Costabel U, Drent M, du Bois R, Grutters JC, Judson MA, Lambiri I, Lower EE, Muller-Quernheim J, Prasse A, Rizzato G, Rottoli P, Spagnolo P, Teirstein A (2011) Defining the clinical outcome status (COS) in sarcoidosis: results of WASOG Task Force. Sarcoidosis Vasc Diffuse Lung Dis 28:56–64

Chciałowski A, Chorostowska-Wynimko J, Fal A, Pawłowicz R, Domagała-Kulawik J (2011) Recommendation of the Polish Respiratory Society for bronchoalveolar lavage (BAL) sampling, processing and analysis methods. Pneumonol Alergol Pol 79 (2):75–89 (Article in Polish)

Cui A, Anhenn O, Theegarten D, Ohshimo S, Bonella F, Sixt SU, Peters J, Sarria R, Guzman J, Costabel U (2010) Angiogenic and angiostatic chemokines in idiopathic pulmonary fibrosis and granulomatous lung disease. Respiration 80:372–378

Déry MA, Michaud MD, Richard DE (2005) Hypoxia-inducible factor 1: regulation by hypoxic and non-hypoxic activators. Int J Biochem Cell Biol 37:535–540

Guérillon C, Bigot N, Pedeux R (2014) The ING tumor suppressor genes: status in human tumors. Cancer Lett 345:1–16

Hunninghake GW, Costabel U, Ando M, Baughman R, Cordier JF, du Bois R, Eklund A, Kitaichi M, Lynch J, Rizzato G, Rose C, Selroos O, Semenzato G, Sharma OP (1999) ATS/ERS/WASOG statement on sarcoidosis. American Thoracic Society/European Respiratory Society/World Association of Sarcoidosis and other Granulomatous Disorders. Sarcoidosis Vasc Diffuse Lung Dis 16:149–173

Iannuzzi MC, Rybicki BA, Teirstein AS (2007) Sarcoidosis. N Engl J Med 357:2153–2165

Kim KK, Kugler MC, Wolters PJ, Robillard L, Galvez MG, Brumwell AN, Sheppard D, Chapman HA (2006) Alveolar epithelial cell mesenchymal transition develops in vivo during pulmonary fibrosis and is regulated by the extracellular matrix. Proc Natl Acad Sci U S A 103:13180–13185

Lazarus A (2009) Sarcoidosis: epidemiology, etiology, pathogenesis, and genetics. Dis Mon 55:649–660

Piotrowski WJ, Kiszałkiewicz J, Pastuszak-Lewandoska D, Antczak A, Górski P, Migdalska-Sęk M, Górski W, Czarnecka KH, Nawrot E, Domańska D, Brzeziańska-Lasota E (2014a) TGF-β and SMADs mRNA expression levels in pulmonary sarcoidosis. Adv Exp Med Biol. doi:10.1007/5584_2014_106

Piotrowski WJ, Młynarski W, Fendler W, Wyka K, Marczak J, Górski P, Antczak A (2014b) Associations between chemokine receptor cxcr3 ligands in bronchoalveolar lavage fluid and radiological pattern, clinical course and prognosis in sarcoidosis. Pol Arch Med Wewn 124:395–402

Sekiya M, Ohwada A, Miura K, Takahashi S, Fukuchi Y (2003) Serum vascular endothelial growth factor as a possible prognostic indicator in sarcoidosis. Lung 181:259–265

Shimoda LA, Semenza GL (2011) HIF and the lung: role of hypoxia-inducible factors in pulmonary development and disease. Am J Respir Crit Care Med 183:152–156

Stenmark KR, Fagan KA, Frid MG (2006) Hypoxia-induced pulmonary vascular remodeling: cellular and molecular mechanisms. Circ Res 99:675–691

Thomas KW, Hunninghake GW (2003) Sarcoidosis. JAMA 289:3300–3303

Tolnay E, Kuhnen C, Voss B, Wiethege T, Muller KM (1998) Expression and localization of vascular endothelial growth factor and its receptor flt in pulmonary sarcoidosis. Virchows Arch 432:61–65

Tzouvelekis A, Ntolios P, Karameris A, Koutsopoulos A, Boglou P, Koulelidis A, Archontogeorgis K, Zacharis G, Drakopanagiotakis F, Steiropoulos P, Anevlavis S, Polychronopoulos V, Mikroulis D, Bouros D (2012) Expression of hypoxia-inducible factor (HIF)-1a-vascular endothelial growth factor (VEGF)-inhibitory growth factor (ING)-4- axis in sarcoidosis patients. BMC Res Notes 5:654

Yamashita M, Mouri T, Niisato M, Kowada K, Kobayashi H, Chiba R, Satoh T, Sugai T, Sawai T, Takahashi T, Yamauchi K (2013) Heterogeneous characteristics of lymphatic microvasculatures associated with pulmonary sarcoid granulomas. Ann Am Thorac Soc 10:90–97

Yu AY, Frid MG, Shimoda LA, Wiener CM, Stenmark K, Semenza GL (1998) Temporal, spatial, and oxygen-regulated expression of hypoxia-inducible factor-1 in the lung. Am J Physiol 275:L818–L826

Zielonka TM, Demkow U, Radzikowska E, Bialas B, Filewska M, Życińska K, Obrowski MH, Kowalski J, Wardyn KA, Skopińska-Różewska E (2010) Angiogenic activity of sera from interstitial lung disease patients in relations to pulmonary function. Eur J Med Res 15(Suppl 2):229–234

Ziora D, Dworniczak S, Niepsuj G, Niepsuj K, Jarosz W, Sielska-Sytek E, Ciekalska K, Oklek K (2000) Proangiogenic cytokines (bFGF and VEGF) in BALF from two different lung segments examined by high resolution computed tomography (HRCT) in patients with sarcoidosis. Pneumonol Alergol Pol 68:120–130

Advs Exp. Medicine, Biology - Neuroscience and Respiration (2015) 15: 71–81
DOI 10.1007/5584_2015_141
© Springer International Publishing Switzerland 2015
Published online: 29 May 2015

Factors Influencing Utilization of Primary Health Care Services in Patients with Chronic Respiratory Diseases

D. Kurpas, M.M. Bujnowska-Fedak, A. Athanasiadou, and B. Mroczek

Abstract

The purpose of our study was to determine the factors affecting the level of services provided in primary health care among patients with chronic respiratory diseases. The study group consisted of 299 adults (median age: 65, min–max: 18–92 years) with mixed chronic respiratory diseases, recruited from patients of 135 general practitioners. In the analysis, in addition to the assessment of the provided medical services, the following were used: Patient Satisfaction Questionnaire, Camberwell Assessment of Needs Short Appraisal Schedule, Acceptance of Illness Scale, and WHO Quality of Life Instrument Short Form. Variables that determined the level of services were the following: age, place of residence, marital status, number of chronic diseases, and level of disease acceptance, quality of life, and health behaviors. The level of provided services correlated with variables such as gender, severity of somatic symptoms, level of satisfied needs, and satisfaction with health care. We concluded that in patients with mixed chronic respiratory diseases a higher level of health care utilization should be expected in younger patients, those living in the countryside, those having a partner, with multimorbidity, a low level of disease acceptance, those satisfied with their current quality of life, with positive mental attitudes, and maintaining health practices.

Keywords

Chronic care model • Chronically ill patient • Family medicine • Family physicians • Primary care • Pulmonary diseases

D. Kurpas (✉)
Department of Family Medicine, Wroclaw Medical University, 1 Syrokomli St., 51-141 Wroclaw, Poland

Opole Medical School, 68 Katowicka St., 45-060 Opole, Poland
e-mail: dkurpas@hotmail.com

M.M. Bujnowska-Fedak and A. Athanasiadou
Department of Family Medicine, Wroclaw Medical University, 1 Syrokomli St., 51-141 Wroclaw, Poland

B. Mroczek
Department of Humanities in Medicine, Faculty of Health Sciences, Pomeranian Medical University, 11 Gen. Chłapowskiego, 70-103 Szczecin, Poland

1   Introduction

The progress in medical science and higher living standards have contributed to the increase life expectancy. Consequently, industrialized countries must face the constant growing challenge to provide health care to patients with chronic diseases. The purposes of healthcare include prevention of complications or deterioration of health, implementation of health needs, and maintenance of autonomy and independence of patients. Chronically ill patients, especially with multimorbidity, benefit from multidisciplinary care in different healthcare institutions, with consequent increase in the risk of mistakes, poor coordination of care, higher healthcare costs as well as low quality of care (Schoen et al. 2014; Coleman et al. 2009). Current global priorities in the care of chronically ill patients indicate the need for development of integrated care, securing the continuity and coordination of care, taking into account the chronic disease burden in the individual and in global dimensions (Bourbeau and van der Palen 2009). Schoen et al. (2014) indicate that the coordination of healthcare becomes flawed when using at least four specialists.

The evaluation of medical services is a component of the quality cycle in healthcare sector (Vedsted and Heje 2008). It allows collecting information concerning the trends in health needs of the population (Sitzia 1999). It is emphasized that the high cost of health care is directly related to the number of highly specialized procedures performed in both outpatient care and during hospitalization. Specialist care increases the costs (Starfield 2008) and in the long term it is not associated with improved quality of care or better clinical outcomes in chronic diseases (Smetana et al. 2007).

Costs of care for the chronically ill account for 78 % of all expenses spent on healthcare in the United States (Hoffman et al. 1996). The most costly procedures include the treatment of: cardiovascular diseases (coronary artery disease, heart failure, and hypertension) and respiratory diseases (chronic obstructive pulmonary disease and asthma), as well as diabetes and psychiatric disorders (Chronic 2004). It is estimated that the care programs for chronically ill coordinating medical services with daily biopsychosocial monitoring might reduce the incidence of hospitalization by 60 % and, using the scale of the United States, decrease costs by $30 billion per year (Chronic 2004).

A meta-analysis on the organization of care for patients with chronic diseases shows that complex programs in accord with the guidelines of the Chronic Care Model improve clinical outcomes and reduce costs of care (Bodenheimer et al. 2002). Expenditures incurred for the treatment of chronic diseases depends on the clinical condition of the patient and the degree of effective prevention of its exacerbation. Treatment of emergencies is more expensive than planned therapy. Usually, it is combined with significant non-medical costs of caring for the chronically ill (Accordini et al. 2006). However, high medical costs stem from the fact that the presence of a chronic disease, in addition to low level of education and a greater use of services beyond primary care, is one of the key indicators of more frequent use of healthcare services, including primary care (Kersnik et al. 2001).

A homogeneous group of patients is usually the main focus of clinical reports. In primary care, however, comorbidity and multimorbidity affect the vast majority of chronically ill patients. As a result, there is a significant lack of reports on the actual level of healthcare services for a heterogeneous group of patients with chronic respiratory diseases in primary care. Consequently, there are no reports specifying factors that determine the level of healthcare services for patients with chronic respiratory diseases. Such data would be useful in the allocation of financial resources in primary healthcare. Therefore, the main purpose of this study was to determine the factors influencing the level of service in primary care for patients with chronic respiratory diseases.

2   Methods

The research was conducted in accordance with the principles of the Declaration of Helsinki. The study was approved by the Bioethical

Commission of the Medical University in Wroclaw, Poland (approval no. KB-608/2011). The main inclusion criteria were: age (at least 18 years) and the diagnosis of at least one respiratory chronic disease.

The study group consisted of 299 adult patients with chronic respiratory diseases. The median age was 65 (min–max: 18–92) years. The study participants were recruited from patients of 135 general practitioners during the period of July 2011–April 2013. The patients who agreed to participate in the project gave written informed consent. The patients were given a questionnaire to complete at home and to return it in a stamped envelope. The information collected concerned the index of healthcare services, the somatic index, and the hospitalization rate.

The index of healthcare services was determined based on the number of identified benefits received during a visit to the doctor. The somatic index was calculated by summing up the values assigned to somatic symptoms and dividing them by 49 (the maximum attainable number of points). Depending on the frequency of somatic symptoms listed by the patient, values from 1 (occurs once a year) to 7 (occurs continuously) were given. The indicator of hospitalization (number of hospitalizations/number of departments) was calculated for each patient for three consecutive years of 2008, 2009, and 2010. The average of these three indicators was adopted as the overall indicator of hospitalization.

Additionally, the Patient Satisfaction Questionnaire, developed by Kurpas et al. (2013a) on the basis of the EUROPEP questionnaire (Wensing et al. 2006), was employed. The questionnaire consists of the following modules: feelings of the patient during the interview and physical examination performed by a physician; information received by the patient from the doctor about the disease, diagnosis and treatment, health promotion and disease prevention; the level of information available by the physician on the patient's disease history (previous visits); providing information to the patient about the disease management (including details on the necessary control visits, specialist consultations, and hospitalization) with the assessment of patient's involvement in the decision making on further procedures; evaluation of the level of communication; concern about the social situation of the patient; attention paid to the patient's emotional difficulties; overall assessment of the family physician, and the evaluation of patient's contacts with non-physician employees of a primary healthcare unit. The answers have been given the following score: 'yes' = 2 points, 'sometimes' = 1 point, 'no' = 0 points. Patient's satisfaction with the services provided was rated by summing up the score. The score range was from 0 to 72 points. The α-Cronbach coefficient describing the internal consistency of this questionnaire is 0.94.

A level of met/unmet needs was estimated with the Camberwell Assessment of Need Short Appraisal Schedule. The questionnaire consists of 24 questions characterizing 22 problematic areas for patients with chronic somatic diseases, excluding mental disorders. To standardize the results the following encoding was used: (1) denoting the state of met needs and (0) denoting the state of unmet needs. The Camberwell index was calculated according to the formula: M/N, where M is the number of needs met and N is the total number of needs. The α-Cronbach coefficient describing the internal consistency of this questionnaire is 0.96.

Quality of life was assessed with the Polish version of the World Health Organization Quality of Life Instrument Short Form (WHOQOL-BREF) within four domains: D1 – Physical, D2 – Psychological, D3 – Social relationships, and D4 – Environmental (Wolowicka and Jaracz 2001). The reliability of the Polish version of the WHOQOL-BREF questionnaire, measured with the α-Cronbach coefficient, refers both to the parts evaluating particular domains (results from 0.81 to 0.69) and the questionnaire as a whole (0.90) (Jaracz et al. 2006).

The patient's adaptation to a life with a disease was assessed using the Acceptance of Illness Scale (AIS) developed by Felton et al. (1984), and adapted to Polish conditions by Juczynski (2009). The AIS consists of eight statements about negative consequences of health state, where every statement is rated on a five-point Likert-type scale (1 denotes poor adaptation to a disease, and 5 its full acceptance). The score for illness acceptance is a sum of all points and can range from 8 to 40. Low scores (0–29) indicate the lack of acceptance and adaptation to a disease and the strong feeling of mental discomfort. High scores (35–40) indicate the acceptance of illness, manifested as the lack of negative emotions associated with a disease. The scale can be used to assess the degree of acceptance of every disease. The α-Cronbach coefficient of the Polish version is 0.85 and that of the original version is 0.82 (Juczynski 2009; Felton et al. 1984).

The author also used the Health Behavior Inventory (HBI) developed by Juczynski (2009). The instrument consists of 24 statements that measure four categories of pro-health behavior: healthy eating habits, preventive behavior, positive mental attitudes, and health practices. The patient marks the frequency related to health behavior and the right activity connected with health on a scale from 1 to 5 (1 – almost never, 5 – almost always). The sum of scores from all four subscales (range 24–120) indicates the general pro-health behavior; the higher the score the healthier behavior. The α-Cronbach coefficient of HBI's internal consistency is 0.85.

2.1   Statistical Analysis

The type of distribution for all variables was determined. The Shapiro-Wilk test was employed to verify normality of these distributions. Arithmetic means, standard deviations, medians, as well as the range of variability (extremes) were calculated for measurable (quantitative) variables, while for qualitative variables, the frequency (percentage) was determined. The analysis of qualitative variables was based on contingency tables and the chi-square test or Fisher's exact test for count data. The Spearman rank test was used to check correlations between pairs of variables.

The analysis of logistic regression was used to examine the influence of explanatory variables on the response variable. This analysis was performed in a number of models using 19 to 6 explanatory variables: sex, having a partner, place of residence, level of illness acceptance, satisfaction with QoL, satisfaction with quality of health state, level of QoL in physical domain, level of QoL in psychological domain, level of QoL in social relationship domain, level of QoL in environmental domain, level of healthy eating habits, level of preventive behaviors, level of positive mental attitudes, level of health practices, age, body mass index (BMI), level of satisfaction from health care, somatic index, and number of chronic diseases. Logistic regression was used to define the influence of explanatory variables on the odds ratio (OR) of health care services index. The chance was defined as the ratio of probability of a certain event to the probability of the opposite event. The chance quotient was defined as the ratio of probability that a certain event happens in one group to the probability that it happens in the other group.

Correspondence analysis was used to provide information similar to the interpretation of the results of factor analysis, but on qualitative variables. Significant differences were defined as the p-value of less than 0.05. The R3.0.2 (for Mac OS X 10.9.4) software was used for data analysis.

3   Results

Detailed sociodemographic data are presented in Table 1, together with diagnoses of chronic respiratory diseases and the most common co-existing diseases. The median number of chronic diseases among the patients was 3 (min–max: 1–15). The median BMI (body mass index) was 27.3 (min–max: 15.6–41.0) kg/m^2 and the median somatic index: 0.4 (min–max: 0.0–1.0). The median index of healthcare services was 5.3 (min–max: 1.0–126.7) and the hospitalization index: 1.3 (min–max: 0.8–5.3).

Table 1 Sociodemographic data of chronically ill patients (n = 299) and their diagnoses

Age (year)		
Q-25 %	55	
Q-50 %	65	
Q-75 %	75	
Min–max	18–92	
Gender	n	%
Women	151	50.8
Men	146	49.2
Place of residence		
Village	131	44.4
Below 5,000	28	9.5
5,000–10,000[a]	14	4.7
10,000–50,000	55	18.6
50,000–100,000	21	7.1
100,000–200,000[a]	21	7.1
Over 200,000	25	8.5
Education		
Primary	79	26.9
Vocational	85	28.9
Secondary	69	23.5
Post-secondary	31	10.5
Higher	30	10.2
Marital status		
Single	34	11.5
Married	188	63.5
Divorced	12	4.1
Widowed	62	20.9
Diagnosis		
Bronchial asthma	111	37.1
Chronic obstructive pulmonary diseases	96	32.1
Unspecified chronic bronchitis	43	14.4
Pulmonary emphysema	39	13.0
Chronic simple and mucous-purulent bronchitis	36	12.0
Bronchiectasis	14	4.7
Most common co-existing diseases[a]		
Primary hypertension	115	38.5
Spondylosis	98	32.8
Atherosclerosis	58	19.4
Osteoarthritis of multiple joints	33	11.0
Non-insulin-dependent diabetes	32	10.7

Diagnoses are presented according to ICD-10

Q quartile

[a]Patients diagnosed as having at least two pathological entities

The results of disease acceptance, assessment of quality of life, health behaviors, and the level of needs met, along with the satisfaction from healthcare provided by the primary healthcare unit are shown in Table 2.

3.1 Associations

The highest index of healthcare services, and thus the highest costs of medical care, was found in men (r = 0.14; p = 0.019), elder patients (r = 0.18; p = 0.002), those having a partner (r = 0.17; p = 0.004), from rural areas (r = 0.18; p = 0.002), with lower level of disease acceptance (r = −0.27; p < 0.0001), low level of satisfaction with QoL (r = −0.23; p = 0.0001), low level of satisfaction with quality of health state (r = −0.27 p <0.0001), low level of QoL in physical domain (r = −0.28; p < 0.0001), low level of QoL in psychological domain (r = −0.20; p = 0.0004), low level of QoL in social relationship domain (r = −0.15; p = 0.010), low level of QoL in environmental domain (r = −0.13; p = 0.027), high level of health practices (r = 0.20; p = 0.007), lower Camberwell index (r = −0.22; p = 0.001), higher level of satisfaction with healthcare (r = 0.22; p = 0.013), higher somatic index (r = 0.28; p < 0.0001), high number of chronic diseases (r = −0.33; p < 0.0001). Statistically significant associations were not found between the index of healthcare services and the level of preventive behavior, healthy eating habits, positive mental attitudes, BMI, and hospitalization index.

Higher BMI was observed in women (r = −0.26; p = 0.001), with higher level of preventive behavior (r = 0.31; p = 0.008), high level of health practices (r = 0.37; p = 0.002), higher somatic index (r = 0.16; p = 0.042), and a higher number of chronic diseases (r = 0.25; p = 0.001). A greater number of chronic diseases was revealed in elder patients (r = 0.40; p = <0.0001), single (r = −0.11;

Table 2 Disease acceptance, quality of life, health behaviors, needs met, and satisfaction from healthcare provided by the primary healthcare unit

Variables	n	Q-25 %	Q-50 %	Q-75 %	Min-max
Level of disease acceptance	282	21.0	26.0	32.0	8.0–40.0
Satisfaction with quality of life	296	3.0	4.0	4.0	1.0–5.0
Satisfaction with health state	294	2.0	3.0	4.0	1.0–5.0
Quality of life					
Physical Domain	296	11.4	13.1	14.9	4.0–19.4
Psychological Domain	297	10.7	12.7	14.7	4.0–19.3
Social Relationships	297	12.0	14.0	16.0	4.0–20.0
Environmental Domain	296	12.4	13.5	14.5	7.5–19.5
Health behaviors					
Healthy eating habit	202	2.8	3.3	3.8	1.0–5.0
Preventive behavior	199	3.2	3.8	4.3	1.0–5.0
Positive mental attitudes	201	3.2	3.7	4.2	1.5–5.0
Health practices	179	3.0	3.5	4.0	1.5–5.0
Camberwell index (level of needs met)	211	0.6	0.8	0.9	0.2–1.0
Level of satisfaction with primary care services	130	49.0	59.0	68.8	19.0–72.0

Q quartile

p = 0.0475), with lower level of disease acceptance (r = −0.33; p < 0.0001), low level of satisfaction with QoL (r = −0.30; p < 0.0001), low level of satisfaction with quality of health state (r = −0.35; p < 0.0001), low level of QoL in physical domain (r = −0.46; p < 0.0001), low level of QoL in psychological domain (r = −0.35; p < 0.0001), low level of QoL in social relationship domain (r = −0.30; p < 0.0001), low level of QoL in environmental domain (r = −0.30; p < 0.0001), high level of health practices (r = 0.24; p = 0.001), higher somatic index (r = 0.48; p = <0.0001), higher index of hospitalizations (r = 0.34; p = 0.021).

3.2 Odds Ratio

The results of logistic regression are shown in Tables 3 and 4. The younger patients (*vs.* older patients) had OR for higher index of healthcare services equal to 4.90. The patients with lower levels of disease acceptance had OR equal to 7.71, with lower levels of satisfaction with QoL equal to 10.70, with higher levels of positive mental attitudes equal to 11.11 and with higher levels of health practices equal to 11.99. Moreover, the patients from rural areas had OR of

4.21, those having a partner had OR of 2.33, and those with the number of chronic diseases of at least 15 had OR of 58.37.

3.3 Correspondence Analysis

The influence of variables on the level of services in the primary care setting is shown in Fig. 1. A higher index of the services provided coexisted with the higher somatic index, older age, a higher number of chronic diseases, lower levels of satisfaction with quality of health state, and with being the resident of rural areas. A weaker relationship was observed between a high level of healthcare services and low levels of disease acceptance, quality of life in all domains, as well as satisfaction with quality of life. A low index of healthcare services coexisted mostly with a low somatic index, younger age, a low number of chronic diseases, and with being the resident of urban areas. There was a weaker relationship of healthcare services with a high level of disease acceptance and a high level of quality of life in all domains. There were no connections to the index of healthcare services in case of such variables as: gender, having a

Table 3 Logistic regression analysis: the response variable is health care services index

i	Variables	Estimate β_i	Std. Error	z value	Pr(>|z|)
Model 1 with 7 variables (n = 183)					
Chi^2 statistic of deviance = 55.9, df = 7, p < 0.00001, pseudo R^2 = 0.22					
1	Having a partner	0.846	0.381	2.222	0.026
2	Place of residence	0.180	0.062	2.894	0.004
3	Level of illness acceptance	−0.064	0.024	−2.668	0.008
4	Satisfaction with QoL	−0.592	0.245	−2.421	0.016
5	Level of positive mental attitudes	0.602	0.245	2.456	0.014
6	Age	−0.021	0.011	−1.965	0.049
7	Number of chronic diseases	0.291	0.088	3.291	0.001
Model 2 with 7 variables (n = 163)					
Chi^2 statistic of deviance = 60.6, df = 7, p < 0.00001, pseudo R^2 = 0.27					
1	Having a partner	1.197	0.429	2.791	0.005
2	Place of residence	0.214	0.072	2.976	0.003
3	Level of illness acceptance	−0.053	0.027	−1.996	0.046
4	Satisfaction with QoL	−0.670	0.268	−2.502	0.012
5	Level of health practices	0.621	0.281	2.210	0.027
6	Age	−0.034	0.014	−2.478	0.013
7	Number of chronic diseases	0.354	0.101	3.506	0.001

QoL quality of life, *Estimate β_i* estimated coefficient β_i in the regression equation, *Std. Error* estimated standard error of coefficient β_i, *z value* value of normal reference distribution, *Pr(>|z|)* two-tailed p-value corresponding to the z value

Table 4 Odds ratios from models of logistic regression for health care services index

Variables	OR	Per unit 95 % CI	1/OR	OR	Per range 95 % CI	1/OR	Range
Model 1							
Having a partner: no (0), yes (1)	2.33	1.12–5.02	0.43	2.33	1.12–5.02	0.43	1
Place of residence: 1–9[a]	1.20	1.06–1.36	0.84	4.21	1.63–11.53	0.24	8
Level of illness acceptance: 8–40	0.94	0.89–0.98	1.07	0.13	0.03–0.56	7.71	32
Satisfaction with QoL: 1–5	0.55	0.34–0.88	1.81	0.09	0.01–0.61	10.70	4
Level of positive mental attitudes: 1–5	1.83	1.14–3.01	0.55	11.11	1.71–81.76	0.09	4
Age: 18–92	0.98	0.96–0.99	1.02	0.20	0.04–0.98	4.90	74
Number of chronic diseases: 1–15	1.34	1.14–1.61	0.75	58.37	5.89–762.96	0.02	14
Model 2							
Level of health practices: 1–5	1.86	1.08–3.28	0.54	11.99	1.38–116.28	0.08	4

OR odds ratio, *CI* confidence interval
[a]Place of residence: city: over 200,000 (1); 100,000–200,000 (2); 50,000–100,000 (3); 20,000–50,000 (4); 10,000–20,000 (5); 5,000–10,000 (6); 2,000–5,000 (7); below 2,000 (8); village (9)

partner, high level of preventive behaviors, and positive mental attitudes.

4   Discussion

The findings of the present study were that healthcare services were determined by patient's age, place of residence, marital status, the number of chronic diseases, disease acceptance, quality of life, and pro-health behaviors. The level of healthcare services was associated with gender, severity of somatic symptoms, the level of needs met, and satisfaction with healthcare. The literature characterizing patients' perception of healthcare services has yielded similar results,

Health care services index (HCS)

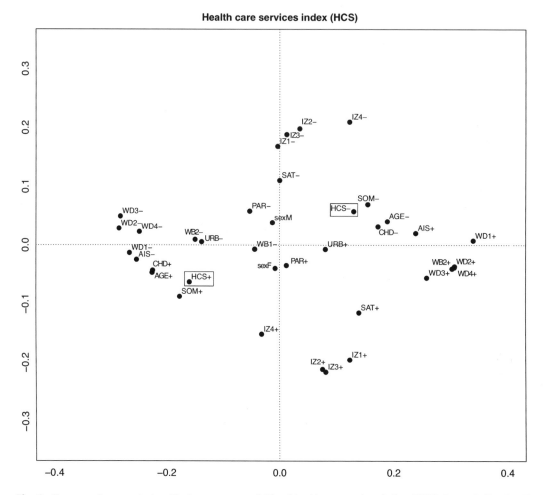

Fig. 1 Correspondence analysis with the response variable of healthcare services index (HCS). Legend: Sex (sexF: female, sexM: male); PAR – Having a partner (−PAR: no, +PAR: yes); URB – Place of residence (−URB: rural area, +URB: urban area); AIS – Level of illness acceptance (−AIS: ≤median, +AIS: >median); WB1 – Satisfaction with QoL (−WB1: ≤median, +WB1: >median); WB2 – Satisfaction with quality of health state (−WB2: ≤median, +WB2: >median); WD1 – Level of QoL in physical domain (−WD1: ≤median, +WD1: >median); WD2 – Level of QoL in psychological domain (−WD2: ≤median, +WD2: >median); WD3 – Level of QoL in social relationship domain (−WD3: ≤median, +WD3: >median); WD4 – Level of QoL in environmental domain (−WD4: ≤median, +WD4: >median); IZ1 – Level of healthy eating habits (−IZ1: ≤median, +IZ1: >median); IZ2 – Level of preventive behaviors (−IZ2: ≤median, +IZ2: >median); IZ3 – Level of positive mental attitudes (−IZ3: ≤median, +IZ3: >median); IZ4 – Level of health practices (−IZ4: ≤median, +IZ4: >median); AGE – Age (−AGE: ≤median, +AGE: >median); SAT – Level of Satisfaction form health care (−SAT: ≤median, +SAT: >median); SOM – Somatic index (−SOM: ≤median, +SOM: >median); CHD – Number of chronic diseases (−CHD: ≤median, +CHD: >median); CAM – Camberwell index (−CAM: ≤median, +CAM: >median); HCS – Health care services index (−HCS: ≤median, +HCS: >median)

concluding that satisfaction with the service provided by a family physician has a major bearing on the number of visits to the doctor and overall impression of medical services (Schoen et al. 2014; Coleman et al. 2009). It has also been assumed that the mere presence of a chronic disease can increase the number of visits to a family physician, which corresponds to the number of services in primary health care (Kersnik et al. 2001).

The present results are in line with those of Glynn et al. (2011) who revealed a greater intensity of medical services among patients with multimorbidity. Those authors found an increased number of consultations in primary care (an average of 11.9 services for patients with 4 or more chronic diseases *vs.* 3.7 in patients without one chronic disease), a greater number of consultations at emergency rooms (3.6 *vs.* 0.6, respectively), and a greater chance of hospitalization (OR 4.51) in multimorbid patients. The overall medical costs in patients with 4 or more chronic diseases were five-fold greater. It has been shown that among patients participating in programs that include multidisciplinary teams, with emphasis on improved communication, the number of hospitalizations and days of readmissions is reduced compared with routine care (Sochalski et al. 2009).

In the present study, we found that a high rate of healthcare services, which translates to high costs of healthcare, was more frequently observed in men, older patients, those having a partner, living in rural areas, and those with low levels of needs met as specified by the Camberwell index. In contrast, the OECD report (2011) has shown that women and people with low income more often report unmet health needs and the lack of necessary care. In that report the most common issues in meeting health needs in the countries of Poland, Finland, and Estonia were the waiting time for medical services and, similarly to Greece and Italy, the costs of treatment.

There are reports pointing to the association of the number of medications used by patients with the number of medical services in primary healthcare (Kurpas et al. 2013b). As stated by Vyas et al. (2012), the consequence of polypharmacy among chronically ill patients is a high risk of side effects, which in turn reduces quality of life and increases the use of medical services. The impact of polypharmacy has been underscored by Kardas (2007) who shows an inverse relationship between the number of medications and the compliance with therapy recommendations, which leads to more frequent visits to a family physician and hospitalizations, thereby increasing the direct costs of healthcare.

A significant problem regarding pharmacotherapy also is a change in treatment during hospitalization, which is not followed by sufficient information to the patient on potential side effects resulting from polypharmacy. That translates into health deterioration during outpatient treatment and increases the frequency of subsequent hospitalization (Schoen et al. 2014). When forming care for the chronically ill elderly patient, attention should be paid to the relationship between the number of physicians prescribing medications and polypharmacy. It is regarded as a marker for the coexistence of several chronic diseases (Sobow 2010). The relationship between polypharmacy and index of healthcare services applies to outpatient care to a greater extent, although it bears significance for the risk of secondary hospitalization as a consequence of pharmacokinetic and pharmacodynamic interactions.

The correspondence analysis used in our study confirmed the results of logistic regression analysis. A high index of healthcare services was found in patients with multimorbidity. Many authors report that the preparation of patients with chronic diseases for self-care, with the accurate information on the control of implementation of physicians and nurses' recommendations, increases the effectiveness of medical interventions and reduces the amount of both visits to emergency units and hospitalizations and hospital readmissions (Bourbeau and van der Palen 2009).

The usefulness of the healthcare services index for the decision making on the allocation of financial resources in primary healthcare needs further evaluation. The presence of a higher healthcare services index in chronically ill patients living in rural areas requires further attention as well. It is worth noting that in medical practices located in rural areas there is a stronger structure of gatekeeping. Because of the distance and suburban communication difficulties, patients may resign from visits to outpatient clinics, often using the advice of a family physician. In the process of strengthening the gatekeeping within the Polish primary care, while improving the quality level, the index of healthcare services will increase. In the long term, this will be associated with reduced costs throughout the whole healthcare system (Starfield and Shi 2002).

5   Conclusions

In patients with mixed chronic respiratory diseases a higher level of health care utilization should be expected in younger patients, those living in the countryside, having a partner, those with multimorbidity, a low level of disease acceptance, those satisfied with their current quality of life, and with positive mental attitudes. The present findings should be taken into account while allocating financial resources for primary health care. They highlight the need for systemic innovation in the area of primary healthcare to improve the quality of life in patients with chronic respiratory system diseases.

Conflicts of Interest The authors declare no conflicts of interest in relation to this article.

References

Accordini S, Bugiani M, Arossa W, Gerzeli S, Marinoni A, Olivieri M, Pirina P, Carrozzi L, Dallari R, De Togni A, de Marco R (2006) Poor control increases the economic cost of asthma: a multicentre population-based study. Int Arch Allergy Immunol 141:189–198

Bodenheimer T, Wagner EH, Grumbach K (2002) Improving primary care for patients with chronic illness. JAMA 288:1775–1779

Bourbeau J, van der Palen J (2009) Promoting effective self-management programmes to improve COPD. Eur Respir J 39:461–463

Chronic care improvement: how medicare transformation can save lives, save money, and stimulate an emerging technology industry. An ITAA E-Health White paper: A Product of the ITAA E-Health Committee, May 2004

Coleman K, Mattke S, Perrault P, Wagner EH (2009) Untangling practice redesign from disease management: how do we best care for the chronically ill? Annu Rev Public Health 30:385–408

Felton BJ, Revenson TA, Hinrichsen GA (1984) Stress and coping in the explanation of psychological adjustment among chronically ill adults. Soc Sci Med 18:889–898

Glynn LG, Valderas JM, Healy P, Burke E, Newell J, Gillespie P, Murphy AW (2011) The prevalence of multimorbidity in primary care and its effect on health care utilization and cost. Fam Pract 28:516–523

Hoffman C, Rice D, Sung HY (1996) Persons with chronic conditions: their prevalence and costs. JAMA 276:1473–1479

Jaracz K, Kalfoss M, Górna K, Baczyk G (2006) Quality of life in Polish respondents: psychometric properties of the Polish WHOQOL-Bref. Scand J Caring Sci 20:251–260

Juczynski Z (2009) Measurement tool in the promotion and health psychology. Laboratory of Psychological Tests, Warsaw, pp 74–80

Kardas P (2007) Compliance, clinical outcome, and quality of life of patients with stable angina pectoris receiving once-daily betaxolol versus twice daily metoprolol: a randomized controlled trial. Vasc Health Risk Manag 3:235–242

Kersnik J, Svab I, Vegnuti M (2001) Frequent attenders in general practice: quality of life, patient satisfaction, use of medical services and GP characteristics. Scand J Prim Health Care 19:174–177

Kurpas D, Church J, Mroczek B, Hans-Wytrychowska A, Nitsch-Osuch A, Kassolik K, Andrzejewski W, Steciwko A (2013a) The quality of primary health care for chronically ill patients: a cross-sectional study. Adv Clin Exp Med 22:501–11.

Kurpas D, Mroczek B, Bielska D (2013b) The correlation between quality of life, acceptance of illness and health behaviors of advanced age patients. Arch Gerontol Geriatr 56:448–456

OECD (2011) Unmet health care needs. In: Health at a glance 2011: OECD indicators. OECD Publishing. Available from: http://dx.doi.org/10.1787/health_glance-2011-52-en. Accessed on 5 Sept 2014

Schoen C, Osborn R, How SKH, Doty MM, Peudh J (2014) In chronic condition: experience of patients with complex health care needs, in eight countries 2008. Health Aff (Millwood) 28:1–16

Sitzia J (1999) How valid and reliable are patient satisfaction data? An analysis of 195 studies. Int J Qual Health Care 11:319–328

Smetana GW, Landon BE, Bindman AB, Burstin H, Davis RB, Tjia J, Rich EC (2007) A comparison of outcomes resulting from generalist vs specialist care for a single discrete medical condition: a systematic review and methodologic critique. Arch Intern Med 167:10–20

Sobow T (2010) Dangers of polypharmacy in neurology. Adv Med Sci 6:483–491

Sochalski J, Jaarsma T, Krumholz HM, Laramee A, John JV, McMurray JJV, Naylor MD, Rich MW, Riegel B, Stewart S (2009) What works in chronic care management: the case of heart failure. Health Aff 28:179–189

Starfield B (2008) The future of primary care: refocusing the system. N Engl J Med 359:2087–2091

Starfield B, Shi L (2002) Policy relevant determinants of health: an international perspective. Health Policy 60:201–218

Vedsted P, Heje HN (2008) Association between patients' recommendation of their GP and their evaluation of the GP. Scand J Prim Health Care 26:228–234

Vyas A, Pan X, Sambamoorthi U (2012) Chronic condition clusters and polypharmacy among adults. Int J Fam Med 2012:193168. doi:10.1155/2012/193168

Wensing M, van den Hombergh P, Akkermans R, van Doremalen J, Grol R (2006) Physician workload in primary care: what is the optimal size of practices? A cross-sectional study. Health Policy 77:260–267

Wolowicka L, Jaracz K (2001) Polish version WHOQOL 100 and WHOQOL Bref. In: Wolowicka L (ed) Quality of life in medical sciences. Poznań, Poznan Medical University Publications, pp 231–238

Advs Exp. Medicine, Biology - Neuroscience and Respiration (2015) 15: 83–89
DOI 10.1007/5584_2015_148
© Springer International Publishing Switzerland 2015
Published online: 29 May 2015

Influence of Iron Overload on Immunosuppressive Therapy in Children with Severe Aplastic Anemia

Katarzyna Pawelec, Małgorzata Salamonowicz,
Anna Panasiuk, Elżbieta Leszczynska,
Maryna Krawczuk-Rybak, Urszula Demkow,
and Michał Matysiak

Abstract

Children with severe aplastic anemia (AA) require multiple transfusions of the red blood cells during the immunosuppressive therapy. This leads to iron overload and manifests as elevated levels of ferritin in blood. The aim of this study was a retrospective analysis of the influence of the elevated serum ferritin on the overall survival, event-free survival, the risk of relapse, and response to treatment in children with AA during immuno-suppressive therapy. We analyzed 38 children with AA (19 girls, 19 boys, aged 2–17 years) treated according to the obligatory protocol for AA in Poland. The response rate was assessed on days 84, 112, and 360. Patients were divided into three groups: group I consisted of children with ferritin below 285 ng/mL (6 children), group II with ferritin between 286 and 1,000 ng/mL (13 children), and group III ferritin >1,000 ng/mL (19 children). Kaplan-Meier plot was used to estimate the overall survival and event-free survival. We found the overall survival did not differ between the three groups. Event-free survival was significantly shorter (p = 0.03) in patients with ferritin levels >1,000 ng/mL compared with the groups with ferritin bellow 1,000 ng/mL. The time to relapse was significantly shorter in group III than in the other two groups (p = 0.02). We also found the differences in the treatment response at day 84 (p = 0.03) and day 112 (p < 0.0001) of immunosuppressive therapy. These findings confirm a negative influence of iron overload in children with AA on the effect of treatment and the risk of relapse.

K. Pawelec (✉), M. Salamonowicz, and M. Matysiak
Department of Pediatric Hematology and Oncology,
Medical University of Warsaw, 24 Marszalkowska St,
00-576 Warsaw, Poland
e-mail: katarzyna.pawelec@litewska.edu.pl

A. Panasiuk, E. Leszczynska, and M. Krawczuk-Rybak
Department of Pediatrics, Oncology and Hematology,
Bialystok Medical University, Bialystok, Poland

U. Demkow
Department of Laboratory Diagnostics and Clinical
Immunology of the Developmental Age,
Medical University of Warsaw, Warsaw, Poland

Keywords

Aplastic anemia • Children • Ferritin • Immunosuppressive therapy • Iron
load • Relapse risk • Survival

1 Introduction

Aplastic anemia (AA) is a rare genetic disease
resulting from bone marrow failure. Immunosuppressive therapy, including cyclosporin and
antithymocyte globulin is used in patients for
whom bone marrow transplantation is not an
option, with response rates of 60–80 %
(Pawelec et al. 2008, 2015; Chang et al. 2010;
Deyell et al. 2011; Bagby et al. 2004). However,
supportive care with red blood cell transfusions
is also essential in many patients to maintain
adequate hemoglobin level (70–80 g/L), particularly since the response to immunosuppressive
therapy may be delayed. Hence iron overload
can be expected in frequently transfused
patients, resulting in organ damage, particularly
liver and heart (Cappellini et al. 2010;
Hoffbrand et al. 2012; Porter et al. 2008;
Takatoku et al. 2007). Transfusion of just one
unit of red blood cells provides about
190–210 mg of additional iron. Iron overload
is one of the major health problems in this group
of patients, as human organism has no regulatory mechanism to discard excess of iron.
The state of chronic iron overload with iron
accumulating in the heart, liver and endocrine
organs ultimately results in significant morbidity and mortality (Romiszewski et al. 2011; Lee
2008; Lee et al. 2010). Iron overload can be
assessed through the measurement of serum ferritin or more invasively through direct determi-
nation of liver iron concentration by means of
biopsy.

The aim of this study was a retrospective
analysis of the influence of elevated serum ferritin on the overall survival, event-free survival,
risk of relapse, and response to treatment in

children with AA during immunosuppressive
therapy.

2 Methods

2.1 Patients

The study was conducted in accordance with the
principles of the Declaration of Helsinki for
Human Experimentation and was approved by
the institutional Ethics Committee. Retrospective
analysis of medical history of 38 patients treated
in 2 Polish centers of pediatric hematology and
oncology in the years 1996–2012 was carried
out. The patients' characteristics are presented
in Table 1. All children fulfilled the diagnostic
criteria of severe aplastic anemia including: bone
marrow hypocellularity, <25 % of the age reference or 25–50 % of the age reference with the
hematopoietic precursors representing less than
30 % of residual cells and the presence of two
peripheral blood abnormalities: (1) absolute neutrophil count <0.5 × 10^9/L, reticulocytes
<20.0 × 10^9/L and (2) platelets <20.0 × 10^9/L
(Camitta 2000).

The patients had no HLA-matched family
donor. The patients' work-up excluded inherited
bone marrow failure disorder and paroxysmal

Table 1 Characteristics of patients with severe aplastic
anemia and iron overload ($n = 38$)

Female/Male	19/19
Age (year)	11 ± 5 (2–17)
Number of transfusion during immunosuppressive therapy	11 ± 5 (6–39)
Level of ferritin (ng/mL)	1,520 ± 1,608 (50–6,148)

Data are means ± SD and range

nocturnal hemoglobinuria. Screening for the latter was carried out using flow cytometry with analysis of CD55 and CD59 expression on neutrophils and red blood cells. The diagnostic procedure toward chromosome instability of Fanconi anemia also was undertaken.

2.2   Protocol

The patients received rabbit-derived antithymocyte globulin (r-ATG) (Lymphoglobulin; Genzyme; Cambridge, MA) at a dose of 3.75 mg/kg, i.v., at days 1–5 and cyclosporine-A (CSA) (Sandimmun Neoral, Novartis Pharma; Nuremberg; Germany) at the dose of 5 mg/kg, p.o., on days 1–180. The dose of CSA was modified according to maintain its serum concentration at 100–200 ng/mL. Granulocyte colony-stimulating factor (G-CSF) (Neupogen; Amgen, Thousand Oaks, CA) was given through subcutaneous or intravenous route only in cases of severe infections non-responding to antibiotics and antifungal medications.

The remission was evaluated on days 112, 180, and 360 since the onset of treatment. Complete remission was defined as the absolute neutrophil count $>1.5 \times 10^9/L$, platelets $>100.0 \times 10^9/L$, and hemoglobin concentration >11.0 g/L. Partial remission was defined as the absolute neutrophil count $>0.5 \times 10^9/L$, platelets $>20.0 \times 10^9/L$, and hemoglobin concentration $>8.0/L$. The overall remission rate was defined as the sum of complete and partial remissions.

2.3   Ferritin

The level of ferritin in the blood was measured by an immunochemical test (Architect Ferritin; Abbott Ireland, Diagnostic Division, Lisnamuck, Longford, Ireland). Normal range of ferritin was defined as 7–285 ng/mL. For the purpose of data analysis, the patients were divided into three groups according to the highest ferritin level during immunosuppressive therapy: (1) six children with ferritin <285 ng/mL, 13 with ferritin between 286 and 1,000 ng/mL, and 19 with ferritin >1,000 ng/mL. Patients with ferritin >1,000 ng/

mL had received deferoxamine (Desferal, Novartis Pharma, Nuremberg, Germany) at a dose of 25 mg/kg, s.c. for 8–12 h three times a week until ferritin dropped below 1,000 ng/mL.

2.4   Statistical Analysis

The overall survival, expressed in years, was defined as the interval between the diagnosis of AA and death of the patient. The event free survival, expressed in years, was defined as the time from the diagnosis to death or disease relapse. In case of no demise or relapse, the time from the diagnosis to the last follow-up with the evaluation of treatment response was considered as the length of event free survival. The longest follow-up period was 10 years. Survival curves were calculated by the Kaplan-Meier method and compared using the Wilcoxon-Gehan test A p-value <0.05 defined statistical significance in all comparisons. Confidence intervals that did not cross 1 were considered significant. Statistical analysis was carried out with Stata 11.0 software (StataCorp LP, College Station, TX).

3   Results

3.1   Survival Analysis

There were six deaths in the study group, including five in the patients with ferritin level >1,000 ng/mL. We failed to detect any influence of the ferritin level on the estimated overall survival and (p = 0.195). AA relapses occurred in three patients who were subsequently qualified for hematopoietic stem cell transplantation from an unrelated donor. A Log-rank test revealed a significantly shorter estimated event free survival in the patients with the ferritin level >1,000 ng/mL in comparison to those with normal ferritin level and ferritin <1,000 ng/mL (p = 0.03) (Fig. 1).

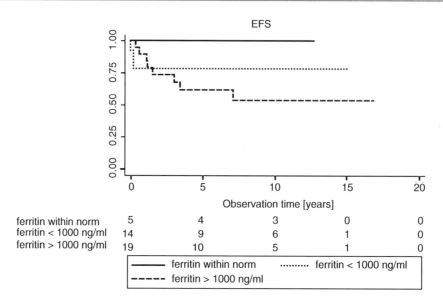

Fig. 1 **Kaplan-Meier event free survival** in patients with severe aplastic anemia according to ferritin level. Normal range of ferritin was defined as 7–285 ng/mL

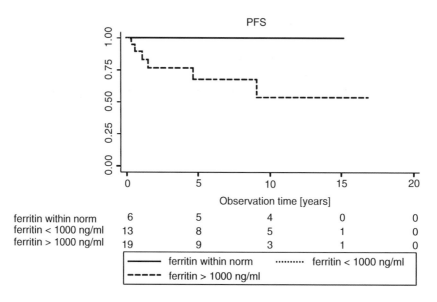

Fig. 2 **Risk of relapse** in patients in relation to ferritin level. Normal range of ferritin was defined as 7–285 ng/mL

Both log-rank and Wilcoxon tests showed a significantly shorter free survival time in the patients with ferritin >1,000 ng/mL in comparison to in comparison to those with normal ferritin level and ferritin <1,000 ng/mL (p = 0.02 for log-rank test, p = 0.04 for Wilcoxon test) (Fig. 2).

3.2   Treatment Response

The response to treatment on day 84 showed seven complete remissions (CR), 14 partial remissions (PR), and no response in 16 patients (Table 2) The patients with the highest ferritin levels >1,000 mg/mL turned out to be

Table 2 The response to the therapy in children with AA on day 84

Ferritin level (ng/mL)	CR (n)	PR (n)	NR (n)
>285	2	2	1
286–1,000	4	6	2
>1,000	1	5	13

Number of cases; *CR* complete remission, *PR* partial remissions, *NR* no response. Normal range of ferritin was defined as 7–285 ng/mL

Table 3 The response to therapy in children with AA on day 112

Ferritin level (ng/mL)	CR (n)	PR (n)	NR (n)
>285	4	0	1
286–1,000	9	1	1
>1,000	2	4	13

Number of cases; *CR* complete remission, *PR* partial remissions, *NR* no response. Normal range of ferritin was defined as 7–285 ng/mL

non-responders more frequently, which was statistically confirmed ($p = 0.031$).

Response to immunosuppressive therapy on day 112 also turned out to be associated with ferritin level. The patients with the highest ferritin level >1,000 ng/mL more often did not respond to treatment in comparison with those having lower ferritin levels ($p < 0.0001$) (Table 3).

4 Discussion

Uncontrolled iron accumulation in patients receiving multiple red blood cell transfusions is a serious problem. Patients with AA treated with immunosuppressive therapy usually require multiple blood transfusions. The Korean Iron Overload Study Group has shown that patients with aplastic anemia or myelodysplastic anemia that iron overload (serum ferritin >1,000 ng/mL) reflects in the number of transfusions and the duration of dependence on transfusions (Lee 2008). The mean ferritin level in the 331 patients of that study, 68 with AA including 58 with severe anemia, amounted to 4,084 ng/mL (range 1,254–22,916 ng/mL). In the present study, severe AA was an inclusion criterion.

The mean ferritin level amounted to 1,520 ng/mL (range 50–6,148 ng/mL). The mean ferritin level on the lower side in our study compared with other studies (Lee 2008) cloud be due to the fact that we included only patients on immunosuppressive therapy, and the ferritin level below 1,000 ng/mL was not an exclusion criterion. Treatment-resistant or relapsed patients were qualified for hematopoietic stem cell transplantation from an unrelated donor. In a Japanese retrospective study conducted in 292 patients with aplastic or myelodysplastic anemia, there were 75 deaths. The ferritin level over 1,000 ng/mL was detected in 97 % of fatal cases (Takatoku et al. 2007). In the present study, six patients died including five with ferritin >1,000 ng/mL. The influence of iron overload on survival of myelodysplastic patients has been investigated in a large Italian retrospective study conducted in 467 patients (Malcovati et al. 2005, 2006). The study has shown that transfusion-dependent patients have a decreased overall survival compared with those who do not depend on transfusions ($p < 0.001$) and has confirmed that iron overload decreases overall survival ($p = 0.003$). Further, a negative effect was more prominent in patients with good prognosis (median survival >100 months) than in those with poor prognosis (median survival of approximately 50 months).

In the present study, we failed to detect any difference in the overall survival between patients with ferritin level >1,000 ng/mL and those with lower ferritin values. However, a significantly shorter event free survival ($p = 0.03$) was observed in the group with ferritin above 1,000 ng/mL in comparison with the other groups. We also showed a relationship between high ferritin level (>1,000 ng/mL) and the time to disease relapse ($p = 0.02$). Taken together, high ferritin level plays a role in survival time and prognosis in patients who undergo multiple transfusions.

Studies of iron chelation therapy in patients with AA have mainly focused on the efficacy in reducing tissue iron and on safety. However, new reports concerning hematologic improvements in association with iron chelation therapy in patients with AA are currently emerging. It is of interest to evaluate hematologic responses in patients with AA receiving iron chelation therapy, as iron overload also has a suppressive effect on erythroid progenitors and may increase transfusion requirements (Koh et al. 2010; Lee et al. 2010, 2013; Oliva et al. 2010). A study on the evaluation of patients undergoing iron chelation with deferasirox assessed the hematologic response taking into account the ferritin level in patients with AA (Lee et al. 2013). That *post hoc* study was conducted in 116 patients from 23 countries with ferritin levels >1,000 mg/mL and the number of transfusions exceeding 20. The response to treatment was analyzed in 72 patients including both severe (n = 9) and non-severe AA (n = 63). Immunosuppressive therapy was applied to seven patients with severe AA and 41 patients with non-severe AA. All patients received chelation therapy with deferasirox. Complete remission was not observed in these patients, partial remission was achieved in 30 patients, including 19 after immunosuppressive therapy. A significant reduction of the initial ferritin level was detected in patients with partial remission. The hematologic response was observed more frequently in the patients who had higher reductions in the serum ferritin, suggesting a relation, at least to an extent, to a reduction in body iron (Lee et al. 2013). Interestingly, non-responders encompassed a higher proportion of patients categorized with severe AA and those who received a higher number of transfusions in the year prior to the entry into the study. It is possible that these non-responding patients constitute a group that would require a longer period of treatment to achieve a hematologic response (Lee et al. 2013; Koh et al. 2010). It is worth mentioning that a significant reduction of ferritin after deferasirox treatment has been described in patients with myelodysplastic, AA, and other rare anemias (Cappellini et al. 2010). Data from the 341 myelodysplastic and 116 AA patients demonstrate that a mean deferasirox dose of 19.2 mg/kg/day was associated with a significant reduction in the serum ferritin (p < 0.05). Many patients with these anemias were chelation-naïve

despite being heavily iron overloaded, which points to the need for a greater awareness of the potential influence of iron overload on the treatment outcome in these groups of patients (Porter et al. 2008). Moreover, iron overload influences the results of hematopoietic stem cell transplantation (Altès et al. 2002; Storey et al. 2009). Patients with severe AA who do not respond to immunosuppressive therapy or have disease relapse qualify for stem cell transplantation.

Storey et al. (2009) have compared the influence of iron overload in both autologous and allogeneic transplants. Using the transplant iron score to define iron overload in both groups, the authors found that allogeneic transplant patients are at a disproportionately high risk of death associated with iron overload. These results suggests that iron overload is a prognostic marker in patients undergoing allogeneic stem cell transplant. The transplant iron score can be considered a predictor of stem cell transplant survival. In view of these observations, special therapeutic programs have been created to decrease iron overload. In Poland, currently, patients with severe AA are qualified to such a program by the Coordination Team for Iron Overload Treatment appointed by the Head of the National Health Fund. Our present observations might also spur increased awareness of the need to monitor iron overload in children with severe AA. We believe that the strength of the present study also lies in having a fairly homogenous group of patients, with the same disease severity and treatment method, in particular that the disease is infrequent. Further follow-up studies are warranted to assess the role of ferritin and iron overload as prognostic factors in severe AA patients.

Conflicts of Interest   The authors declare no conflicts of interest in relation to this article.

References

Altès A, Remacha AF, Sureda A, Martino R, Briones J, Canals C, Brunet S, Sierra J, Gimferrer E (2002) Iron overload might increase transplant-related mortality in haematopoietic stem cell transplantation. Bone Marrow Transplant 29:987–989

Bagby GC, Lipton JM, Sloand EM, Schiffer CA (2004) Marrow failure. In: Hematology ASH Education Book 1:318–336

Camitta BM (2000) What is the definition of cure for aplastic anemia? Acta Haematol 103:16–18

Cappellini MD, Porter J, El-Beshlawy A, Li CK, Seymour JF, Elalfy M, Gattermann N, Giraudier S, Lee JW, Chan LL, Lin KH, Rose C, Taher A, Thein SL, Viprakasit V, Habr D, Domokos G, Roubert B, Kattamis A on behalf of the EPIC study investigators A (2010) Tailoring iron chelation by iron intake and serum ferritin: the prospective EPIC study of deferasirox in 1744 patients with transfusion-dependent anemias. Haematologica 95:557–566

Chang MH, Kim KH, Kim HS, Jun HJ, Kim DH, Jang JH, Kim K, Jung CW (2010) Predictors of response to immunosuppressive therapy with antithymocyte globulin and cyclosporine and prognostic factors for survival in patients with severe aplastic anemia. Eur J Haematol 82:154–159

Deyell RJ, Shereck EB, Milner RA, Schultz KR (2011) Immunosuppressive therapy without hematopoietic growth factor exposure in pediatric acquired aplastic anemia. J Pediatr Hematol Oncol 28(6):469–478

Hoffbrand AV, Taher A, Cappellini MD (2012) How I treat transfusional iron overload. Blood 120:3657–3669

Koh KN, Park M, Kim BE, Im HJ, Seo JJ (2010) Restoration of hematopoiesis after iron chelation therapy with deferasirox in 2 children with severe aplastic anemia. J Pediatr Hematol Oncol 32:611–614

Lee JW (2008) Iron chelation therapy in the myelodysplastic syndromes and aplastic anemia: a review of experience in South Korea. Int J Hematol 88:16–23

Lee JW, Yoon SS, Shen ZX, Ganser A, Hsu HC, Habr D, Domokos G, Roubert B, Porter JB (2010) Iron chelation therapy with deferasirox in patients with aplastic anemia: a subgroup analysis of 116 patients from the EPIC trial. Blood 116:2448–2454

Lee JW, Yoon SS, Shen ZX, Ganser A, Hsu HC, El-Ali A, Habr D, Martin N, Porter JB (2013) Hematologic responses in patients with aplastic anemia treated with deferasirox: a post hoc analysis from the EPIC study. Haematologica 98:1045–1048

Malcovati L, Porta MG, Pascutto C, Invernizzi R, Boni M, Travaglino E, Passamonti F, Arcaini L, Maffioli M, Bernasconi P, Lazzarino M, Cazzola M (2005) Prognostic factors and life expectancy in myelodysplastic syndromes classified according to WHO criteria: a basis for clinical decision making. J Clin Oncol 23:7594–7603

Malcovati L, Porta MG, Cazzola M (2006) Predicting survival and leukemic evolution in patients with myelodysplastic syndrome. Haematologica 91:1588–1590

Oliva EN, Ronco F, Marino A, Alati C, Pratico G, Nobile F (2010) Iron chelation therapy associated with improvement of hematopoiesis in transfusion-dependent patients. Transfusion 50:1568–1570

Pawelec K, Matysiak M, Niewiadomska E, Rokicka-Milewska R, Kowalczyk J, Stefaniak J, Balwierz W, Załecka-Czerpko E, Chybicka A, Szmyd K, Sońta-Jakimczyk D, Bubała H, Krauze A, Wysocki M, Kurylak A, Wachowiak J, Grund G, Młynarski W, Bulas M, Krawczuk-Rybak M, Leszczyńska E, Urasiński T, Peregud-Pogorzelski J, Balcerska A, Wlazłowski M (2008) Results of immunosuppressive therapy in children with severe aplastic anaemia. Report of the Polish Paediatric Haematology Group. Med Wieku Rozwoj 12:1092–1097

Pawelec K, Salamonowicz M, Panasiuk A, Demkow U, Kowalczyk J, Balwierz W, Zaleska-Czepko E, Chybicka A, Szmyd K, Szczepanski T, Bubala H, Wysocki M, Kurylak A, Wachowiak J, Szpecht D, Młynarski W, Bulas M, Krawczuk-Rybak M, Leszczynska E, Urasinski T, Peregud-Pogorzelski J, Balcerska A, Kaczorowska-Hac B, Matysiak M (2015) First-Line immunosuppressive treatment in children with aplastic anemia: rabbit antithymocyte globulin. Adv Exp Med Biol 836:55–62

Porter J, Galanello R, Saglio G, Neufeld EJ, Vichinsky E, Cappellini MD, Olivieri N, Piga A, Cunningham MJ, Soulières D, Gattermann N, Tchernia G, Maertens J, Giardina P, Kwiatkowski J, Quarta G, Jeng M, Forni GL, Stadler M, Cario H, Debusscher L, Della Porta M, Cazzola M, Greenberg P, Alimena G, Rabault B, Gathmann I, Ford JM, Alberti D, Rose C (2008) Relative response of patients with myelodysplastic syndromes and other transfusion-dependent anaemias to deferasirox (ICL670): a 1-yr prospective study. Eur J Haematol 80:168–176

Romiszewski M, Gołębiowska-Staroszczyk S, Adamowicz-Salach A, Siwicka A, Matysiak M (2011) The use of chelation therapy in the treatment of iron overload in a girl with Diamond-Blackfan anemia – a case report. Nowa Pediatr 4:90–94

Storey JA, Connor RF, Lewis ZT, Hurd D, Pomper G, Keung YK, Grover M, Lovato J, Torti SV, Torti FM, Molnár I (2009) The transplant iron score as a predictor of stem cell transplant survival. J Hematol Oncol 2 (44):1–9

Takatoku M, Uchiyama T, Okamoto S, Kanakura Y, Sawada K, Tomonaga M, Nakao S, Nakahata T, Harada M, Murate T, Ozawa K, Japanese National Research Group on Idiopathic Bone Marrow Failure Syndromes (2007) Retrospective nationwide survey of Japanese patients with transfusion-dependent MDS and aplastic anemia highlights the negative influence of iron overload on morbidity/mortality. Eur J Haematol 78:487–494

Advs Exp. Medicine, Biology - Neuroscience and Respiration (2015) 15: 91–92
DOI 10.1007/5584_2015
© Springer International Publishing Switzerland 2015

Index

A
Animal model, 51–58
Aplastic anemia (AA), 83–88
Artificial intelligence, 9
Aspiration, 15–22, 52
Aspiration syndrome, 51–58

B
Beige adipose tissue, 26–29
Biomarkers, 4, 6, 31, 63
Bronchoalveolar lavage fluid (BALF), 2–5, 46, 54, 55, 57, 63–68

C
Chest tube drainage, 15–22
Children, 25–31, 37, 83–88
Chronically ill patient, 72, 75, 79
Chronic care model, 72
Computer modeling, 9

D
Diagnostic, 7–12, 44, 63, 84, 85
Disease marker, 1–6
Disease mechanisms, 2, 4, 63, 67

F
Family medicine, 71
Family physicians, 73, 78, 79
Ferritin, 84–88
Fibroblast growth factor 21 (FGF21), 25–31
Finite elements method, 7–12

H
Heat shock proteins (HSPs), 42, 45–48

I
Immune regulation, 35
Immunosuppressive therapy, 83–88
In silico analysis, 8, 9, 12
Insulin, 5, 26, 29–31, 35–39

I
Interleukin-18 (IL-18), 2–5
Interleukin-33 (IL-33), 1–6
Irisin, 25–31
Iron load, 83–88

L
Leptin, 27, 30, 31, 35, 37–38
Lung fibrosis, 2, 5, 62, 68
Lung injury, 53, 57, 58

M
Manual aspiration, 16–22
Meconium, 51–58
Mycobacterium, 41–48

N
Nitrate/nitrite, 42, 44, 45

O
Obesity, 8, 25–31, 35–39

P
Peripheral blood mononuclear cells (PBMCs) culture, 43–47
Peroxynitrite, 41–48
Pleural space, 16, 17
Pneumothorax, 7–12, 15–22
Primary care, 71–80
Prognosis, 62, 87
Pulmonary diseases, 1–6, 16, 43, 61–68, 72

R
Radiological classification, 66
Regulatory T cells (Tregs), 35–39, 42
Relapse risk, 84, 86
Respiratory support, 52, 53

S
Sarcoidosis, 1–6, 16, 41–48, 61–68
Small bore catheter, 15–22

Spontaneous pneumothorax, 15–22
Survival, 42, 84–88

T
Tuberculosis (TB), 16, 42–48

V
Ventilation, 8, 51–58

W
White adipose tissue, 25–31